"THANK YOU MY DUCK!"

"THANK YOU MY DUCK!"

Mabel Warner

The day-to-day story of a
BIRD AND ANIMAL SANCTUARY
created by an ordinary working
Oxfordshire Woman
slightly eccentric maybe!

This true story is about the creation of a dream being put into reality — not an easy task by any means.

I would hesitate to advise another person to do the same on such a limited budget, it's become a bottomless shaft, but I'll die a very happy person having achieved my aim.

Rima Publishing

Rima Publishing
The Old Chapel, Milcombe, Banbury, Oxfordshire OX15 4RP

First published 1998

© Mabel Warner 1998

All rights reserved. No part of this publication may be reproduced in any material form including photocopying or storing in any medium by electronic means and whether or not transiently or incidentally to some other use of this publication) without the written permission of the copyright holder. Applications for the copyright holder's written permission to reproduce any part of this publication should be addressed to the publishers and the copyright holder.

British Library Cataloguing in Publication Data
A catalogue record for this book is available from the British Library

Library of Congress Cataloguing in Publication Data
A catalogue record for this book is available from the Library of Congress

ISBN 0 9532915 0 2   Hardback
ISBN 0 9532915 1 0   Paperback

Manuscript typed by Jane Gillett, Chipping Norton
and edited by Joyce Holt, Adderbury
Origination, further editing and composition by
Richard Payne, Paragon Print & Design, Oxfordshire, England
Printed and bound in Great Britain

To all animal-lovers

**This book is all about
LOVING, CARING and SHARING
together with animals and birds.**

TITLE:

"My duck" is a well-known expression in Oxfordshire, particularly among older country folk. Nothing else as a title for this book would have summarised so aptly the deep love and affection I feel for my native county as the phrase I use so often at the Waterfowl Sanctuary.

FOREWORD

I first met Mabel Warner in 1988 when I dropped into her florist shop to order a wreath. We started chatting and it wasn't long before she was telling me of her plans for a Water Fowl Sanctuary. Had it been anyone else, I may well have dismissed her story and thought she was a crank, but there was a quiet determination in her voice that made me think that here was a woman who would not give up until she achieved her ambition.

Like an acorn into an oak tree, the Sanctuary grows and matures daily. It is loved by the many visitors, especially the children who are encouraged to pick up and stroke the small animals in the *"cuddle corner"*.

Mabel always looks more comfortable surrounded by her beloved birds and animals than she ever does out of that environment amongst strange humans. The growing success of the sanctuary is due solely to her sheer dedication and hard work and I admire her greatly. I wish I had her energy and zest.

I have visited the sanctuary many times: each time it has grown a little more and there is something new to see. At the rate she is going, it soon won't be *"Mabel's little Sanctuary"*. I wish her everything that she wishes for the sanctuary and hope it continues to flourish.

You go in a stranger, but you come out a friend.

Councillor Wendy Humphries
Chair of Cherwell District Council

ACKNOWLEDGEMENTS

In particular, I would like to thank Pauline and Gary, Richard and Mary, Joyce and John, John Simms and John Rogers for their help, but also very many more friends, too numerous to mention by name.

God bless you all!

I would be pleased to think that the profit from the sale of this book will help towards the care and maintenance of birds and animals.

On the Taylor side of my family my grandmother Ellen Deely came from the great landowner Deely family who owned land from the Midlands to the coast. It is said that one Jabus Deely was an overseer in the days of the enclosures. No doubt he was not very popular with many folk as he drove around in his pony and trap, a portly little figure with a watch and chain hanging on his chest. My brother Christopher heard from our Dad, Charles Taylor, that two men came from Oxford bearing a piece of paper for him to sign. In return for a relatively small amount of money he relinquished all claims to ground in which he apparently had some shared family interest. This land is, in fact, the site in central Oxford on which the Taylorian Institute and the Ashmolean Museum now stand. Very few assets passed down to my grandparents in spite of the fact that they both stemmed from wealthy families. Her marriage to our Grampy, Jonas Taylor, was not well received by her family. I remember him, a lovely "gentleman of the Lord", (a chapel preacher) climbing on to his soap box decrying all the evil that existed in the world.

Grampy Jonas Taylor

The Taylors farmed at Arncott, around the *Tally-ho* pub. Granny Taylor, née Deely, worked hard at her farm shop at the old Homestead, Fencott, with its lovely bow-window. From these memories my recurring dream was that I too should have a shop, something I realised later for 25 years.

During the 2nd. World War, our Dad went to work in the Morris car factory, Oxford, as money was so short, especially in small farming. At that time he gave a lift to Andrew Franklin, whose family still live in the cottage opposite the Church in Charlton-on-Otmoor, my place of birth.

My Dad, Charles, was given one acre of land, including the site of a former cottage, on which he built, for £120, a four-bedroomed timber, steel and tin bungalow. The whole family helped dig an air-raid shelter beneath it in 1939. After the war, my Dad bought out his brothers' and sisters' shares in the Homestead Farm opposite, where I lived until I was married, together with my three brothers and three sisters. After letting it for some years for evacuees, he eventually sold the bungalow to the sitting tenants.

Jonas at the Homestead on a happy tea-party with the Sunday School children

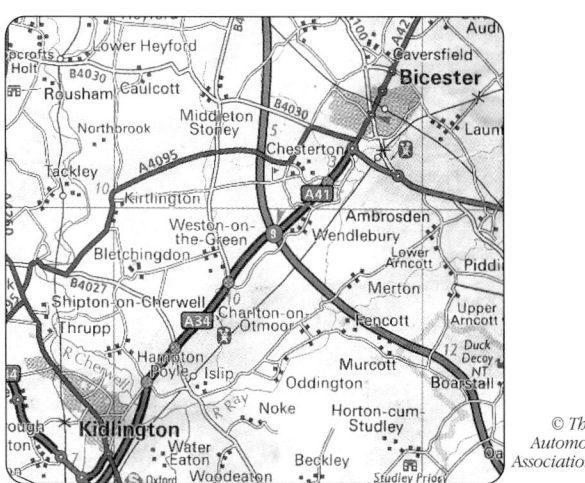

© The Automobile Association 1995

Myself at the Homestead Farm at Fencott

Evelyn took this snap of the family: Dad, Mum, Peggy, John, Mabel, Raymond, May & Christopher

Cyril and Peggy's Wedding — 24th May 1947
Mum, left, Dad, right; Fred North behind; Grampy North second from right; Audrey, Evelyn, Olive, Mabel and May; in front: Sue and Christopher.

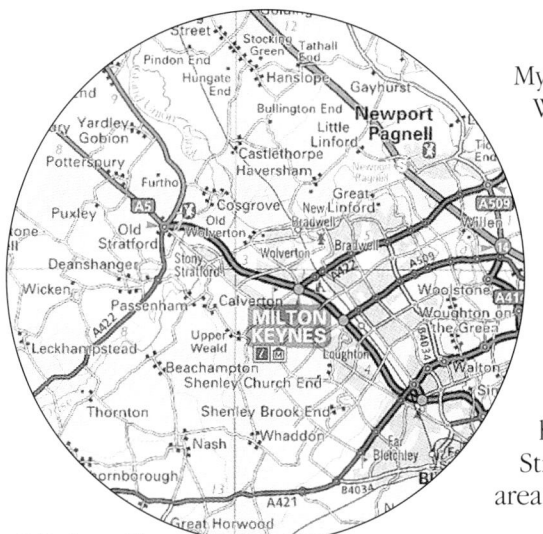

© The Automobile
Association 1997

My mother Mary Agnes Dickens, was the niece of William Dickens, a hay-dealer to the Government in the 1st. World War (1914-1918). He made a fortune supplying hay for the horses of the Cavalry in both England and France. While William enjoyed great wealth, and was astute in buying up many properties in Wolverton, Deanshanger and Stoney Stratford, his own brother, Joseph Dickens, (Mary's father and my grandfather), on the other hand, was employed as a poorly paid labourer (hay-tier) renting one of his many cottages at Cosgrove Farms near Stoney Stratford. William's properties extended over an area which is now a large part of Milton Keynes.

My grandmother, Georgina Dickens, died of cancer when my mother was just twelve years old. She was in charge of four younger sisters for several years. Her only little brother, Joseph, was killed by climbing a damson tree, on to a hay rick, falling and breaking his neck. His father picked him up by his head and clicked his neck back into place and he was sent to school. Mary had to leave her classroom and take little Joey home as he was nearly dying; his death followed shortly after. Her sister, Eva, also died of the influenza epidemic that was sweeping the country at the time. My mother was then taken to London to the Dickens' household at Barnes in order to escape the fate of her sister. She was treated by a Harley Street doctor and remembered being transported by the Dickens' carriage and pair. Later she recovered and was given singing lessons, since she had a beautiful voice. While in London, an incident took place which made her flee the Dickens' household to a Young Ladies' Christian Mission where she signed a pledge never to drink, smoke or swear. The four sisters received a mere £2000 each in 1965. The Dickens aunties owned a lace-making and dress-making shop in London and trained many young ladies in the art of lace-making. Some of them made lace for our present Queen Elizabeth the Queen Mother.

Charles & Mary

Mum & Dad's Wedding – Grandpa Dickens behind bride; Grampy Jonas Taylor & wife, Ellen, seated on left; Uncle Sid, left of groom. Also in picture: Agnes, May, Gladys, Mabel and Lucy. Seated to the right: Mr & Mrs Hatwell and Daisy On the ground: Bernard Treadwell and William North

I was born at Charlton-on-Otmoor, Oxfordshire on April 8th 1931, being taken at three months to live at Fencott and then to the Homestead Farm opposite. My father, Charles Taylor, had a hard time bringing up seven children, of whom I was the middle one. Our mother, Mary, (née Dickens), born in 1901, was a patient, gentle, Christian lady, who never said an unkind word about anyone, nor had she drunk alcohol, smoked or worn make-up. She was Queen Mum to all her children, grandchildren and great-grandchildren.

Mum with little brother Chris

School started for me at three and a half at Charlton-on-Otmoor and I went on to Gosford Hill, Kidlington. At school each day up to the age of fourteen I just lived to get home and out on the farm. Mucking out calves, pigs etc. were everyday tasks. Rain, snow or wind never bothered me. I dug the garden and mowed the lawns around the homestead — all good training for the future. On the farm there were always ducks, geese, hens, pigs and cows. A special cow was a strawberry Roan called Dinky, a part of our family and provider of milk and butter through the latter years of the Second World War. She was milked by myself and most of my brothers and sisters. The more we loved her, the more milk she gave us.

After leaving school I had to go to Elliston & Cavell in Oxford as a seamstress, earning just 19s 11d per week, half of which went on bus fares, half to mum.

William (Bill) Robbins, with the help of his teacher Bessie Johnson, gained a scholarship to Lord William's Grammar School at Thame. We married at Murcott Chapel on January 6th 1951 and moved to Horton-cum-Studley. For five years we lived at Hilltop Cottage near Studley Priory, where William worked as a gardener. Captain Henderson allowed us to fence and use a field behind the cottage, where we kept a calf and a pig to generate income to buy our first acre of land from Cyril Crawford for £80.

Our best friend at that time was Jack Badger, one of the largest landowners in the village, and with a heart to match. He would stop his tractor and unload one bale of hay for our cow. He also allowed our cow and her calf to graze in his field with his herd without charge because he recognised an ambitious and hard-working young couple. Thank you, Jack.

When our first daughter was born in 1956, we had a bungalow built by Jack Durndell, costing £1,400, on an acre of land and created our first nursery at Horton-cum-Studley, opposite Warren Farm. From this beginning we progressed into the horticultural and floristry trades and had florists' shops for 25 years, 15 of which were *'Interflora'*, in Oxford, Lechlade and Banbury.

Our First Bungalow

For many years my husband and I and our three children visited the Domestic Fowl Trust, Slimbridge Wildfowl Trust, Folly Farm, The Rare Breeds Survival Centre and other collections of animals and birds around the country. The children always had pets; family photographs were invariably taken cuddling one of them.

Bill with Polly & cat

Bill loved all gentle creatures. For some thirty years he kept a Blue Amazon parrot, (not gentle with anyone else other than himself). From accounts of older village folk who remembered its history, we believe it was nearing one hundred years old when it died, two months before our son was born in 1970. One year later to the day of its death very uncanny red claw marks appeared on Bill's arm exactly where the bird used to sit every evening. He wept when that bird died. I have always believed that it had to die to make room for our son, as the bird used to be very jealous of our daughters if their father picked them up.

From 1956 until his death in 1984 Bill and I were always self-employed, but, as he was a great worrier, I do not think he would have gambled on this dream of mine.

Myself with Swans 1950

Aerial view of The Nursery Horton-cum-Studley 1959

Jackie with pigs 1958

Jackie & Wendy with a row of Bill's sweetpeas

This "dream of mine" was only made possible after Bill died. In 1985 I married my childhood sweetheart, Dennis Warner, who, being a generous and unselfish man, allowed me to be myself. *(See page 97)*

In June 1987 he saw an advertisement in the paper for 22 acres of very neglected wetland, facing south, located up a quiet country road at Wigginton Heath, near Hook Norton, soon to be offered for sale by auction. Following the sale of my farmhouse and 12 acres at Farnborough, Oxfordshire, and moving into Dennis's small bungalow, at Church Enstone, I walked the Wigginton land with a view to its purchase, together with my eldest daughter, Jackie, and her husband, Anthony. Having studied survey maps which had pinpointed a spring in one corner of the top fields, we located a source supplying the land. (Since buying it, we have found three more.)

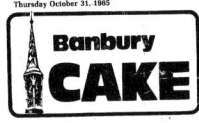

After an assurance from my bank manager, I went to the White Lion Hotel in Banbury to bid. I managed to buy three fields in three lots for £27,000, a reasonable price for grade 6 agricultural land which was running with water and quite useless for anything else, but perfect for what I had in mind. It seemed within my limits for setting up a Waterfowl Sanctuary, a long time ambition. Trifleton Waterfowl Collection, near Crosshands, Wales, had long impressed me with its natural running water and ponds, backed by springs, and this is exactly what I envisaged for Wigginton Heath.

Investing most of my capital into this venture, I firstly purchased a dismantled ex-army timber building (60ft x 20ft). My son-in-law Anthony took me to Blackpool to view and buy it. At £600 it was a very reasonable price to pay in 1987 for a building of this size. We arranged transport, which cost another £400, and unloaded it temporarily at Rectory Farm, Church Enstone. I advertised locally for a contractor to move and erect the building in the bottom field at Wigginton Heath as a rearing and feed shed. At the same time I bought a second-hand, steel-framed agricultural barn, which was erected by Andy Gilkes opposite the timber shed, making a useful and attractive entrance to the Sanctuary.

All these developments attracted quite a bit of interest in the nearby village of Wigginton. Letters to the district council resulted in a visit, followed by a letter to my home enquiring what I intended doing on the land. I was able to assure an officer that I intended to make good use of otherwise useless land, so providing leisure enjoyment for many people.

Thank you my duck!

JCB starts work

The Knightsbridge Hotel in Banbury was then being demolished, and I started buying in stone and brick rubble to go into the front yard. Tons and tons kept sinking, but it eventually hit the clay bottom and hardened up after lorries had firmed it down.

Roy Invine from Chipping Norton came to do all the woodwork, making huts for ducks and geese and putting up cattle-proof fences around the perimeter. Dennis and my son, Rodney, helped each day. Most of the timber used to make the huts was purchased very cheaply from Banbury Demolition and Eynsham Park Sawmills. Martin at Bruern Woodlands estate office supplied rough natural timber off-cuts which were used round the front yard. These blended in naturally with this kind of environment.

I contracted a man with a JCB to dig out the shallow ponds (one was already in existence) at the far top end of the three acre field. He dug out three large ponds and one small one. Then Martin Stevens from Chipping Norton put his machine to work and things began to take shape. He made his machine work wonders creating natural-looking ponds linked by small winding streams.

We then had the top field drained bringing water down to, and through the ponds and streams, soon filling up to give the effect I wanted. The lads dammed up most of the ponds at the outlets with timber to raise the level of the water. The ponds had solid clay bottoms, similar to those at Otmoor, where I was raised, so I believed that they would hold water without using costly liners.

The Barn nearing completion: Den & myself in a pause for reflection

Roy Invine & Den: yet more construction work

The old faithful David Brown tractor

Hard work on the layout

A grass mixture was planted on the spread soil around most of the ponds. The next job was to build fences around each individual pond and to link up electric fencing top and bottom. We used three-foot wire netting with straining wires top and bottom. Each enclosure is entered by a gateway of rustic timber and each has a hut in which birds can shelter at night if they so choose.

On the other side of the three acre field, Roy, Den and Rod built an enclosure with six-foot high wire fencing embedded eight inches into the ground to prevent foxes digging underneath. It was then divided into thirty separate runs, each with a house for the chickens, bantams, turkeys, guinea-fowl, peacocks, pheasants and some smaller species of duck and geese. Each had a separate gate made with rustic wood.

The goldfish pond at the left hand top corner was fitted with a close mesh outlet running through a pipe down to a small open ditch at the end of each fowl pen. This works well in winter months while springs are running, but our first year (1989) open to the public was the driest summer since 1976. Therefore, two of our shallower ponds dried up. However, when rain came they soon filled up again, flowing well and tumbling down gently on and out of the bottom through the old culvert beneath the disused railway line.

As far as we know, our topmost field where the springs emerge has never been sprayed with chemicals. Completely untouched by modern farming a wealth of flora and fauna is found here along and around the nature trail we are encouraging our visitors to explore. Little owls and tawny owls live here and a pair of kestrels raised three young here in 1989. Sparrowhawks have also been seen along with dozens of other species of butterflies and birds.

The lads created an observation tower. Roy arranged to purchase on my behalf from Bruern Abbey four tree trunks fifteen feet high and sixteen smaller ones. Then, after my sketch was altered until it was completely satisfactory, the tower was built. It is an exceptionally sturdy construction, quite a work of art and I was extremely pleased with it. It has since been traversed by many thousands of feet and appreciated by one and all, for the view over the ponds and surrounding countryside is stunning.

Rabbits started to arrive: the circumstances being those of unwanted pets, owners moving house, students leaving home, children going to boarding schools, etc. Next, I decided to make a rabbit warren outside, to enable rabbits to dig and burrow in the earth. It has since proved very popular with little visitors, and many lovely comments have been written in the visitors' book about it.

From left: *Nigel Andrews, Maurice North, Roy Invine and myself prior to Opening Day*

Anticipation

The lead-up to the opening of the Sanctuary at Easter 1989 was a very exciting time for us all. After placing a series of advertisements in local newspapers, the media started taking an interest in what was being created at Wigginton Heath.

A letter arrived from the local council asking me to tell them what I was planning for the agricultural site. I immediately went to the office and invited the officers to come and inspect it. They did, and were pleasantly surprised and very interested in what had been achieved so far. But as I had laid down tarmacadam on the paths surrounding the ponds to enable wheelchairs to be taken along, the officer told me that now I would have to obtain planning permission as this action turns agricultural land into a leisure facility. Planning was applied for and approved, allowing me to continue for five years, making it necessary to re-apply providing it is established as an amenity necessary for the public. I feel sure it is so far.

Goat pens were then built in the top left-hand corner of the field. First to come to the Sanctuary was Poppy, a golden Guernsey goat brought by Jane from Marston St. Lawrence. Poppy is a shaggy coated, lovable animal, who, being deaf and barren, had been attacked by the rest of the herd. They had bitten the fur from her ears making her quite a sad spectacle. Now Jane and her family visit regularly and are delighted with her recovery and extremely happy nature in our environment. A pygmy goat called Thomas was introduced next, but unfortunately he likes to butt people, so had to be tethered. Two British Saanans were bought from a neighbour. I now know why he wanted to get rid of them. One in particular would never stay in her pen despite the warm, comfortable house with a hay rack built for her. We have since given sanctuary to three more character goats. They take up many hours of our time and our visitors love them all dearly. I have also bought a Soay sheep who was a frightened, nervous wreck when she arrived. She calmed down after I fed her human stress tablets and I have since found her a husband. This is the start of our own flock of Soay, the oldest breed of sheep and ancestors of all modern sheep. I foolishly imagined that I could create flower-bedecked ponds, but with so many ducks and geese they soon took their toll of anything but the hardiest of waterside plants.

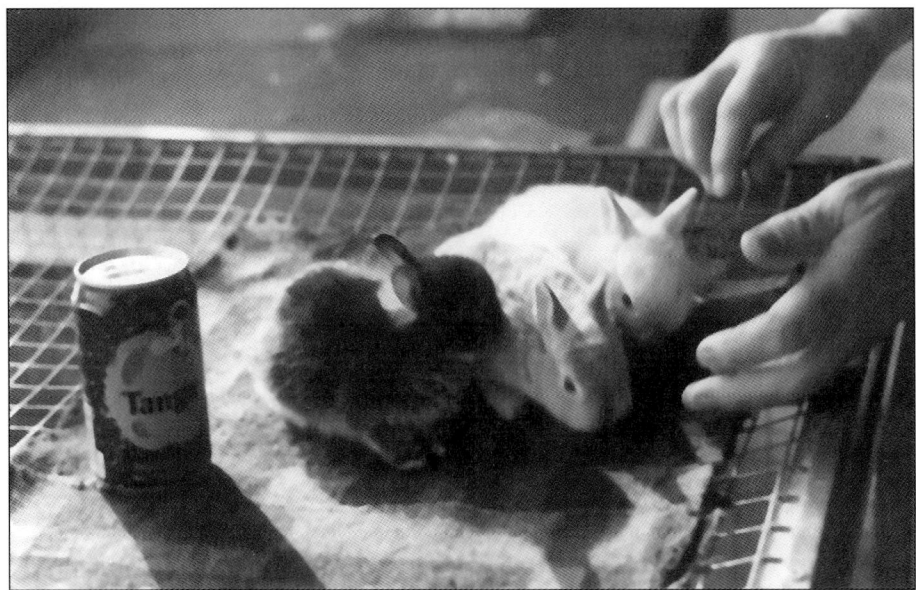

Mabel's Midgets, six weeks old and ready to leave Mum, just the size of a Tango can

Rabbits Galore!

I have loved rabbits since childhood and seem to have an ability to calm them down from vicious bad-tempered, frightened creatures to lovable pleasant natured rabbits. One in particular, a black one called George was about to be put to sleep by a vet as the owner's children could no longer play with him, or handle and love him as rabbits should be loved. A neighbour told them about the Sanctuary and after a telephone call they brought him along to me. He was a beautiful rabbit indeed, with fear-filled enormous eyes, but after showing him love and continuous stroking, talking and singing gently to him he soon became quite friendly and manageable. I fed him stress tablets for humans which I bought from a chemist, so those and my own form therapy worked miracles.

Another rabbit called Reabie, a brown and white buck, arrived. He had been neutered and he, too, understandably was quite bad-tempered. The same therapy treatment was given to him with the same happy results.

Another brown doe came to live with us, we call her Squirrel as she has no ears. We were told she lost them as a youngster but were not enlightened as to how. Nonetheless she is a very good mother, loves having babies and will escape from just about anywhere to meet up with a prospective mate.

The Angoras are another story. They came to me in a very tangled state via RSPCA. It takes around three hours to comb, clip and disentangle the fur when it has been neglected. I now have eight in my care. Baby Angoras are about the sweetest-looking animals of all. My rabbits are happiest outside where visiting children go amongst them, feed them and play with them.

B.W.A.

An Association of Enthusiasts interested in Keeping, Breeding and Conserving Waterfowl.

After joining the British Waterfowl Association (BWA) and reading the many interesting articles and buying guides we visited a breeder, Brian Boning, in Norfolk to purchase our pair of black swans. He went to a great deal of trouble to obtain an unrelated pair for breeding, we hope, in the future. They are magnificent. They live on the island pond queening over all the other occupants but never bullying, just gracefully waiting their turn at feeding time allowing little ducks to have their fill first. They make a beautiful singing sound, music to my ears, quite different to the rather loud cackling and screeching of farmyard geese.

Luckily the Sanctuary is well away from neighbours. Our nearest neighbours keep poultry themselves so our delightful sounds are no worry to them.

A pair of peacocks was given me as a wedding present in 1985 by my brother, John, and his wife, Eileen. Unfortunately, we lost one male before it reached maturity. We then bought a replacement male which died of heart failure also before maturity,— leaving one female. After paying a visit to a gentleman on the Welsh Border who had no female peacocks we managed to supply him with two young females (donated by my kind brother John) in exchange for a mature cock bird. He immediately fell in love with our lone female and as time progressed they produced seven youngsters, but once again we had bad luck. Our dreaded enemy, the stoat, came and killed all the babies. We shall persevere.

Percy in full display

The goldfish pond is looking very healthy with many fish being bred and the large ones growing larger. The tiny silver and gold Koi carp are surviving well.

The day before we left on a holiday in Israel with my sister Peggy, one of my mother Muscovy ducks came out of the yard with twelve ducklings. I wonder if anyone else has had hatching so late in the season. They all survived under my son's care, but I missed the first week of their lives. However, as time goes by I shall breed many more Muscovies as they are one of the most prolific breeds of duck. The first ducks I purchased were Muscovies, which, to my mind, have quite attractive brilliant red head and neck markings, and make excellent mothers.

Peggy and myself in Jerusalem, October 1988

My sister, Peggy North, is loved by everyone who meets her.

She is an artist in her own right, and as children we both loved making fairy gardens and waterfalls in our little stream at Fencott.

I remember making clay ducks and other models that we would dry in the sun.

Her pottery and her paintings of flower fairies reflect our happy childhood days.

We spent an unforgettable holiday in Israel together, travelling by coach around the Holy Land.

She is a Methodist lay preacher, like Grampy Taylor.

Quackers and Quackling

First Residents

The first two ducks given to the Sanctuary came from Norfolk. My friend June's sister had kept them in her greenhouse for years. One was black and white, one white, one Quackers, one Quackling.

However, one day before we had the outside gates fitted someone or something lured Quackers away. We do not think it was a fox on this occasion, as there were no feathers. I was very saddened by this loss, but Quackers had laid two eggs. We were able to hatch them beneath a bantam and had another duck just like her mother.

At the beginning of November 1988 breeder Nick Baughan purchased a pair of rare Reaves pheasants on my behalf. He opened the lid and out flew the hen bird, up and away out of the back of my florist's shop in Calthorpe Street, Banbury. Nick promised to bring me a replacement the following week, but I immediately rushed round to the newspaper office and inserted an advertisement offering a reward for the return of the bird. A lady rang to say she had sighted a bird answering this description. Then another lady rang to say the bird was in her garden looking lost and lonely. I asked her if she could leave her garden shed open and try to lure the bird in with food. She had already gone to great lengths to find out to whom the bird belonged. About 6.30 pm my son Rodney and I visited the house in Bloxham Road. After searching the garden we spotted the pheasant perching on top of a six foot fence. Rodney crept silently towards her and with a swift movement caught her before she could fly away, placing her in a carrying case. I was very pleased indeed to have saved a rare breed from the many perils of nature. I gave the lady, Mrs Wrench, a £10 reward and she immediately handed it back to help the Sanctuary funds. This money went towards purchasing our first three grey Chinese geese and they were named after the children of Mr. and Mrs. Wrench, Emma, Joanne and Jemma.

In November 1988, Hulls of Long Compton drained the back fields and dug a deep ditch along the bottom of the three acres taking the excess water from the seven acre field into the culvert and then beneath the old railway embankment. They revealed evidence of old clay drainage pipes that had lain beneath the surface, broken and blocked solid with earth, therefore quite useless.

A benefactor Mrs. Andrews of Balscote, heard about the Sanctuary and offered the contents of her fish pond, as she was moving. Along went the men and removed the plants and fish. I put them all into my new fish pond. They soon became well established and one year later the goldfish pond is a sheer delight. I could spend many happy hours just gazing at them. Frog spawn has been added, resulting in many small frogs being sighted around the Sanctuary, a good proportion escaping the ducks' puggling beaks. My friend Joyce Soaffe also brought water reeds, yellow bog iris etc. from her wet ground near Otmoor. They are establishing well.

As I have several electricity poles crossing my land I imagined it would be quite an inexpensive operation to obtain a supply to the buildings but after negotiation a price of £4,000 was agreed to install it. Another pole was erected in the roadside hedge to carry a transformer. After my perimeter road was laid it had to be dug up once again to lay the cable. The telephone was a lot less trouble to obtain.

My brother-in-law Maurice, and Darren North built three toilets including one for the disabled and buried a two thousand gallon septic tank beneath the ground. They concreted the floor of the barn and put concrete block walls and large windows all around the reception barn. Glazier Chris from Banbury and Buckingham Glass came to glaze the windows.

I went to the council to tell them we were nearly ready to open to the public on Good Friday, March 24th 1989. I asked if I could put up small white cut-out ducks with arrows indicating the way to the Sanctuary, as we lie well away from the main road. Alec from Banbury made the duck cut outs. The answer I received was ambivalent. *"If you put them up they may be cleared away by the council; on the other hand they may decide not to"*. One year later approximately half still remain intact, the other half being removed by vandals and gales.

On recommendation from the council I applied for official tourist board signs and was told I might get them in several months. I was delighted that they feel I am worthy of signs for my effort and dedication to give joy to others as well as myself. They were erected one year after opening.

MARCH 24TH 1989

My friend Joyce Holt on opening day

The opening day arrived, the weather was fine although a small marquee had been erected on the lawn area in case of rain. The Mayor of Banbury, Councillor Ray Buckley, came to open the Sanctuary. Our other chief guests were Mr. Alan Birkbeck and Miss Clarke, British Waterfowl Association past president, and committee members. Visitors started to arrive and the car park soon filled up, overflowed into the back field and down the side of the road. Everyone looked happy with what they saw, which was very satisfying after all the thought and effort that went into this creation. It was a wonderful, wonderful day.

The Cheshire Home brought a young man called Andrew who is unable to sit up but, lying on a chair-bed, he was wheeled round the duck pens. His face was a picture of appreciation of a very rare occasion for him, just one of the very grateful disabled visitors for whom the Sanctuary was partly created.

Marion and Michael Hart and family brought along two pet lambs they had reared by hand. Caroline and Alan Muddiman brought an exhibition of rare breeds of sheep and one baby pet lamb, circulating it round the visitors. It proved very popular.

A beautiful photograph of my black swans was published in the Ark Magazine of Rare Breeds. After a P.R. firm had distributed press releases and advertisements announcing the opening of the new Sanctuary, several newspapers picked up my story. A reporter from the *Oxford Mail* came for an interview accompanied by a photographer. They were quite impressed, especially by Pinny, a miniature Appleyard Duck, born crippled and unable to stand when first hatched. A visitor brought her to the Sanctuary, as she seemed to be full of spirit and sheer guts, and she turned out to be a real survivor. We called her Pinny as when she tried to walk her feet turned inward, pin-toed; (interbred, I fear).

Thank you my duck!

Banbury Guardian 30th March 1989

Banbury Mayor, Cllr. Ray Buckley with Mabel and Susie the Sebastopol Goose at the Opening.

Opens Good Friday, 24th March 1989

Sanctuary has 800 residents

Maurice, Nigel, Roy and myself

Population growth, especially around larger towns and cities and the onslaught of modern agriculture are just two of the new developments which have had a profound effect on the countryside.

Hedges and woods; even ponds are disappearing at an alarming rate.

Those who enjoy birds and the natural world in general are at last coming to realise that a land rich in wildlife is a precious heritage which we can no longer continue to take for granted.

An Oxfordshire family group led by conservationist Mabel Warner and her farmer husband Dennis, assisted by Roy Irvine, the craftsman behind all of the rustic timber work, is reacquainting people with the pleasures to be gained from the birds and other creatures by establishing a waterfowl sanctuary at Wigginton Heath, not far from Banbury.

The sanctuary has its grand opening on Good Friday at 2 pm by the Mayor of Banbury Coun Ray Buckley and the past president of the British Waterfowl Association Alan Birkbeck and will remain open thereafter daily from 10.30 a.m.

Morning comes early for the 800 or so residents of the waterfowl sanctuary. Everyone from the Warners' smallest Hottentot Teal duck to the big black African turkeys has a job.

There is food to be fed (and watch) and cleaning and preening to be done in order to look attractive for the visitors who they hope will be pouring in once the sanctuary is open. There must be no open gates, no broken fences.

Guard rails and other safety devices will be in place. Public toilets, including the special facility for the disabled, must be clean and fresh.

Visitors who know their birds will recognise most of the different species. Others will have no trouble as easily read signs will help to identify them all.

Thirty-five pens and 15 ponds are home to European shell-ducks, decoy mallard call ducks, red-crested pochard and little diving ducks. Alongside are chestnut-breasted teal, Buff Orpingtons, blue Swedish and black Indian runners, which stand upright like a penguin.

FARMYARD

There are black and white magpie ducks, Emden geese and Greylags (the ancestors, says Mabel, of all domestic geese except the Chinese, and they have a pair of them).

The familiar farmyard Muscovy is there and the Rouen, which attains weights of around eight to ten pounds.

More pens house strutting little bantams and Guinea fowl which make, says Mabel, excellent 'guard fowl' as their piercing shriek instantly warns when people are about. White, bronze and black turkeys gobble importantly from their snug retreat.

The Australian black swan is now a protected species with exports controlled to specialist breeders.

A large aviary houses smaller birds such as cockatiels, parrots, canaries, budgies, finches and several types of pheasant.

While the entire sanctuary is designed to appeal to all the family, certain parts are geared directly at the younger set. A "Pet's Corner" allows kiddies to go right in among the friendly, gentle lambs, goats, rabbits and guinea pigs.

Poppy, a lovely goat who is not only deaf but barren is sure to be a special favourite with her good temper and soft eyes as will be a small Soay sheep, again the original ancestor of all domestic sheep.

A rustic observation tower allows a good overview of all pens and ponds.

There is even more to see and do on the 22 acre site which surrounds the actual sanctuary.

A 'nature trail', fully signposted, leads around the perimeter of these fields while a 'dipping pond' with safety rails allows schoolchildren to take back water samples to explore the contents of the pond.

Falconry displays will take place periodically.

Admission is £1 for adults and 50 pence for children and OAPs. For obvious reasons dogs are not allowed.

Banbury Guardian 23rd March 1989

Back in January I purchased from Trevor Lay of Waveney Wildfowl a pair of European shelducks with very beautiful colouring, a pair of red breasted pochard, (amusing looking ducks), and one pair of tufted ducks (entertaining ducks). They came by rail safely to add colour and variety to the island pond, the largest pond of the sixteen which had been dug.

Moorhens began to visit shyly. One night one had walked across the wet cement in front of the Reception door leaving a delightful semi circular pattern for posterity.

The Reception Barn

A visitor told me that a goat was in the reception barn eating feed from a bucket. I promptly enticed the goat back into the pen which we had laboriously created for her but which was not high enough. Goats jump like deer when they take it into their minds to do so. This particular goat feels the need to be near people, as she came from Mr. Hughes, our neighbour, who kept her close to the farmhouse.

When a photographer came to take photos of the animals, Thomas, a small-horned goat, piebald in colour, decided to eat the case belonging to the poor chap. Apparently, having been reared among teenage boys, who encouraged him to dance and prance using his horned head to its best advantage in the skirmishes, he now thinks everyone should play rough with him. He must be separated from visitors but the remainder of the goat herd are quite gentle and love to be patted and fed.

A fair number of pigeons with damaged wings are regularly brought into the Sanctuary for care. They are very soon able to fly freely again.

Tony Baldry our local M.P. brought his family. He was very impressed and pleased with my creation. He discussed providing replacement ducks for his home village pond at Wroxton.

Many newspapers and magazines bombard me for advertising at this time, to my great cost. It is quite frightening when the bills are waiting to be paid. Some advertising helps but a lot is quite unnecessary.

Second Honeymoon

My first honeymoon was spent working so my second honeymoon was a leisurely time spent in St. Lucia, West Indies. As I love birds, the obvious choice of a hotel was *"The Green Parrot"* where I mistakenly imagined there would be many beautiful green and coloured parrots in the surrounding trees. I was sadly informed by the proprietor that most of the parrots had been killed for food by the native occupants of the beautiful tropical paradise island. However, there were many tiny humming birds darting and glistening in the sunshine. One unfortunate tiny bird flew into the large plate glass window and was killed instantly. Chef Harry, King of *The Green Parrot* hotel, took this tiny bird in his huge black hands and placed it at my request in his deep freeze ready to take back when I returned to England, to be preserved. When asked if anyone practised taxidermy on the island, his answer was, *"It would be stolen in no time, just turn your back and hey presto, it would be gone."*

Thank you my duck!

Whilst there, visiting one of the many out-tracks, I discovered a deep pool between the rocks full of beautiful pink conch shells. Of course I wanted to take back as many as I could carry to England and carried five in my picnic bag. I asked the hotel taxi driver to return to the same spot the next day to gather up some more. I promised him £10 for that favour, possibly way over what I believed to be his likely scale of income. When I saw him every evening I asked him if he had been and gathered up the shells for me; each day his reply was *"tomorrow I will go."* When the last evening came my husband told me that he had paid him £10 NOT to get them. I was exasperated but had to laugh when he said he was not going to carry sea shells home to England. I did not see the taxi driver again so I will never know if Den really did pay him not to get them!

END OF MARCH 1989

My friend Joyce Holt helped by taking visitors' entrance money for me whilst I was still working in my little flower shop in Calthorpe Street, Banbury. I am torn between my love of flowers and the longing to be outside at my beloved Sanctuary.

Here begins my "Glorified (and inconsistent) Diary"!

MARCH 30TH 1989

Two Cayuga ducks arrived, large jet black domestic birds with beetle green sheen, quite old ones. Nice write-ups in the *Banbury Guardian* and *Banbury Cake*.

MARCH 31ST 1989

Joyce helped again, approximately 200 visitors. Three young silver pheasants brought into care.

APRIL 1ST 1989

Up at 3 am. to Birmingham Market. Bought plants, flowers etc. for Sanctuary and shop – a long day. My dear sister May came with cakes etc. to sell in catering caravan; also, among others, Darren Hancox, from Duns Tew, a really nice young man, came to help serve the teas.

My bank manager, Andy McLellan came with his wife and children — very pleased with it all. Approximately 400 visitors today and the children love the baby ducks and chicks.

Maurice, and Darren North finished the cedar wood cladding on front wall of the reception barn. This had been a cedar tree, planked and cured. I had bought the timber several years ago in an antique barn in Warwickshire. I have always loved wood, especially pine. When cut, the aroma is beautiful and the finished product perfection in my eyes.

The *'Love a Duck'* badges sell well at 25p; also pencils, stickers etc. Many well-wishers, friends and helpers today.

APRIL 2ND 1989

Glorious weather for our opening week. Sunday, breakfast in bed made as usual by my ever loving husband, four tiny bantam eggs quite delicious but I am sure they are rare breeds I am eating, (perish the thought!). I do have rather too many of my favourite Columbian Rose Crown bantams who are such happy little chickens. Always first out of the house a.m. and last to go to bed p.m. busy foraging all day long. Similar to Light Sussex in colouring, they are quite different in nature, with attractive rose crowns. I also have a trio of large Light Sussex hens at this moment.

APRIL 3RD 1989

Roy off this week – felling small larch trees to make an adventure playground in back field. Many seats placed around everywhere, made very hurriedly by Roy and friends prior to opening.

APRIL 5TH 1989

I sold a trio of Campbells to Ben Licence, fourteen years old. A very keen lad.

APRIL 6TH/10TH 1989

5000 visitors already! A visitor brought a tub full of horse chestnut seedlings. I planted them out but, like many others, they perished in the drought of the summer.

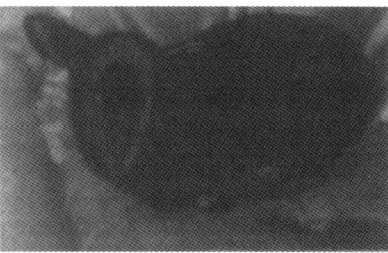

Dorothy Invine's pet lamb

APRIL 15TH 1989

Four geese now sitting on eggs, one Muscovy duck also. The Lord has been good to us guiding many visitors here. For worried and distressed people, it is therapy itself to come to the Sanctuary . They seem to soak up the peace and tranquillity of the beautiful countryside and many hundreds of happy birds and animals. The majority of visitors go away looking much more relaxed and happy than when they arrived. A very inexpensive therapy treatment.

APRIL 16TH 1989

A lovely drawing came by post from a little visitor from Buckingham, first of many, the special one though being the first. The Penhurst school children from Chipping Norton, followed their visits by sending lovely drawings of the birds and animals also 'thank you' letters. Bless their little hearts, it is so pleasing to give them the chance to see God's living creatures at close hand, to hold chicks, ducks and baby bunnies.

Penny Vine

Penny Vine from Radio Oxford came to record a chat with me and the sounds of geese, ducks, turkeys etc. to be broadcast on May 1st, Bank Holiday Monday.

A visitor brought a Campbell drake named Trevor. I put him in a pen and he promptly flew away up and over to the nearest stream the other side of the disused railway. Off went the whole family in search of him. They returned about half an hour later looking quite sad and afraid Mr.

Fox would make a meal of him at night. After they left, this same Trevor came waddling into the car park. Unfortunately the visitors did not leave their names or telephone numbers so I was unable to tell them the good news. I think they said they were from Brailes.

I bought a pair of Kakarekes £45 – very beautiful birds, dark green colour. The male died two days later, the female lays many eggs, unfertile, of course. So sad, she needs a mate.

A family bought newts for my fish pond, rescued from a garden pond being filled in. They were snatched quite literally from beneath the JCB shovel — yet another pond to disappear beneath cement and stone.

APRIL 26TH 1989

A cold misty morning. I took down a pair of ducks to Adderbury lakes and was met at 8.30 am by Joyce Holt. The village magazine *Adderbury Contact,* of which she is editor, is celebrating ten years of publication and the editorial committee decided to make a suitable commemoration of this event in their gift of waterfowl.

APRIL 28TH 1989

Weather good, quite a few visitors. There is an interest in buying young small ducks, seems I could sell small ones when I breed any as they are suitable for small gardens and small ponds.

MAY 1ST 1989

My interview with *Radio Oxford* sounded good. Takings very good over the Bank Holiday weekend. Bloxham School visited — they seem to love the Sanctuary.

MAY 2ND 1989

Five baby Sebastopol geese hatched — very excited about that as they are quite rare, the original pantomime goose. They first came into this country from Russia in the early 1920s, I understand. One baby disappeared, so I removed the whole family into an aviary for safety. Later I lost the mother to a fox, — very sad, but father goose protected his babies and reared them to maturity.

MAY 6TH 1989

About one hundred and fifty visitors today – the visitors' book proves very interesting. Finished netting over the pigeons' pen. Let them out to fly free: much better. They have to be kept in the aviaries for approximately six weeks to establish their home territory.

A teacher from Kineton, Bernard Smith, borrowed my black stag turkey to mate with his hen bird. Very pleased with results when he returned him to me.

A visitor brought a tiny Mallard duckling only approximately twenty-four hours old found abandoned in Aston. I introduced it to the others being reared under an infra-red lamp and it grew up alongside the domestic ducks quite happily. Another visitor rang to say she had found a baby duckling in a park in Oxford, no others or mother to be seen anywhere. She brought it over to Wigginton. At that time a small blue Pekin bantam had just been disappointed with her clutch of eggs. None was fertile so I gave this little duckling to her to mother. She treated it as her own and it grew and grew until it was actually bigger than her but still it tried to nestle beneath her. She would never leave her side until it had to go on the pond but still kept returning to her, getting as close to the edge as possible and continuing to talk to it, she never fell in.

MAY 8TH 1989

A teacher from St. John's School, Banbury, Mrs. Williams, came with eleven-year-olds; she is quite impressed, and will visit later with more pupils.

May 9th 1989

Some buff cochins and partridge cochins were brought in: beautiful birds with feathered legs, looking as though they are wearing trousers.

May 11th 1989

Twenty-two children came from Great Tew playgroup and enjoyed it. Muscovy ducks arrived in Yorkshire safely, thanks to John Holt from Adderbury, Joyce's husband, who took them when returning to his York office. They were collected at a cross roads.

I need higher fences for goats. Trouble, trouble, trouble – they are such lovable animals, forever wanting to be near humans.

Official Council Sanction

May 18th 1989

Cherwell official sanction of Sanctuary for five years licensed to trade, subject to certain conditions. A visit had been made by the whole committee prior to the meeting at Bodicote House. Joyce and I attended, it was 7 pm by the time my application was reached, number thirty one. All through the day the committee members had sat around the table looking quite straight, some slightly bored and very tired, but when they reached my application it was announced *"Ah, now for the highlight of the day, the Waterfowl Sanctuary, Wigginton Heath."* A great change took place, every face around that great table glowed with pleasure and smiles, smiles, smiles.

Sebastopol Goose and young

May 20th 1989

A lovely photograph of my Sebastopol geese in the newspaper. Ducks sitting everywhere now. Polly Parrot died: the vet thought she was nearing one hundred. She had been with us for twelve years, rescued from near death, as she had belonged to an old lady who went into hospital care. The bird would mime a person crying and sobbing, (I assume she copied the old lady); also she barked like a dog and crowed like a cockerel. The bird was quite grey and we thought her to be an old grey, but after a few weeks of correct feeding with fresh fruit etc. her true colours began to show through and it became obvious that she was a Blue Amazon. Parrots must be the most resilient of all birds to be able to survive in that state then go on to live another twelve years. She is now in deep freeze along with a collection of other birds etc. as one day it is my intention to find a taxidermist to work on them. Shortage of money at this time prohibits me. I have a beautiful white barn owl picked up on the roadway going up to Birmingham market at 3.30 am one morning. It was still warm and must have flown into a passing lorry. I have a snow goose killed by a fox, also a little black East Indian duck.

Thank you my duck!

MAY 25TH 1989

It seems I am not able to keep small ducks very successfully as they fall prey to wretched stoats who attack and kill them by biting their throats, despite electric fencing. More thought must go into ways of keeping these little pests out. To date they have killed Lyson teal, chestnut breasted teal, white calls, mallard calls, pintails, mandarin and Carolinas. One pair of miniature Appleyards disappeared but I put those down to a two-legged predator on firm suspicions. A coach party today, our first, the driver just managing to get into the car park. There were also a full mini bus from Social Services.

MAY 27TH 1989

One little boy named Craig had spent the whole of his life in hospital, a hole-in-the-heart patient. His very first visit out was to us. The little boy had never seen anything like it: he was thrilled with all we have to offer here. It is a joy to me to see little faces beam with smiles and wonder at my ever growing family of feathers and fur.

JUNE 1ST 1989

My dear friend Joyce Holt launched the *"Adopt a Duck"* scheme, after receiving a telephone call from Mrs. Bubb of Bodicote who wanted an unusual present for her little granddaughter, Madison Anne Fowler, to be christened in St. Paul's Cathedral. Joyce designed this duck card below, then went to the printers to get copies made. We followed up with *"Adopt a Rabbit"*, *"a Goat"* and *"a Swan"* for £5 to Sanctuary funds.

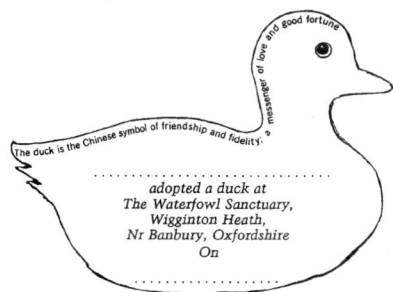

A visitor came to reception asking *"Is the Muscovy duck really supposed to be in your goldfish pond?"* No. A hasty dash up to the pond and removed the offending duck before it consumed every goldfish and my frog spawn. Luckily some had hatched into small frogs but I am sure some of them ended up inside my naughty duck. Clip, clip, clip of flight feathers to prevent that little episode happening again.

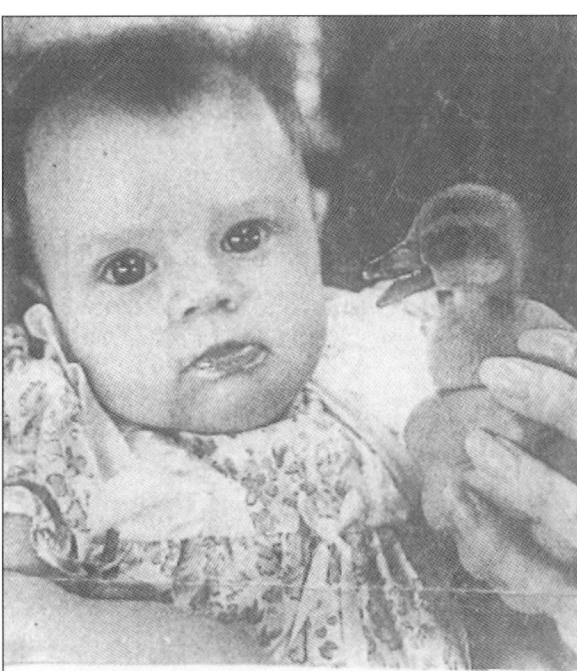

Four-month-old Bodicote toddler, Madison Ann Fowler, was presented with an unusual christening present... a Muscovy duckling from Wigginton waterfowl sanctuary.

Florence Bubb, one of Madison's neighbours at the Red House, Keysar Road, decided to take up the recently opened sanctuary's "adopt a duck" scheme as the perfect present for a little girl called Fowler.

In return for a small fee the scheme offers the ideal present for the conservationist or animal lover — the chance to sponsor a fowl of their own.

HIGHLIGHTS

For Madison, pictured above with her fluffy present, it was one of the highlights of a memorable christening. The toddler — named after the mermaid from the hit film Splash — was christened in St Paul's Cathedral: an honour earned because her grandfather Ian Garner was awarded the MBE for his civil service work. Her great-uncle, Canon Geoffrey Holley took the service.

The "adopt-a-duck" scheme got another customer this week. Amanda and Dan Wright received a duck called Jack as a wedding present on Saturday.

Bird & Animal Sound Recordings for the Blind

JUNE 3RD 1989

A group of visitors from Cheltenham came to record the many different sounds heard around the Sanctuary for the Listening Books for the Blind. It is a good contribution to unfortunate people not able to come to see, but who can listen in their own homes. They sent me a copy tape.

Roy made a lovely chipmunk cage for these delightful little creatures (4' x 4' x 8' high) and they entertained everybody with their amusing actions, never keeping still, dashing up and down, round and round continuously until evening when they retired to their nesting box to sleep. Chipmunks are from North America and from Siberia. A member of the squirrel family, they are prettily marked with brown stripes to the body with fluffy, flat tails; they measure about eight inches long from nose to tail, eat nuts, fruit and seeds and enjoy burrowing in deep bedding.

JUNE 4TH 1989

I was told I had the *"best job in the world"* by schoolteacher John Ekers of Bloxham College, who visits regularly with his children.

JUNE 5TH 1989

Baby cockatiels found in a nest box. They hiss like little snakes when approached.

Quick action today averted another tragedy. A commotion was coming from the pen of the eight week-old ducklings and I saw a stoat disappearing rapidly through the fence. A little magpie duck had a slashed neck, (just missing the jugular vein), but it survived in the warmth of the aviary. In future I must be more sensible in heating aviaries, as my electricity bill is frighteningly high.

JUNE 7TH 1989

First Easington brownies visited, twenty-one children and ten adults, working for badges. A very nice group, they thoroughly enjoyed their visit.

JUNE 26TH 1989

Following a meeting with my accountant and my bank manager I was advised to stop spending now on major projects and to manage my finances better.

Lorraine Carter and children frequent visitors from Cutteslowe

JULY 1ST 1989

Mother peahen came off her nest with five babies, but once again I did not like to disturb her so left it to nature. Nature is so cruel and I have lost them all to stoats once again. I've learned my lesson now not to leave any baby young stock outside, even under netted roofs.

JULY 2ND 1989

Muscovy duck brought off a brood of ten babies, two blue-grey. I'll keep my blue to try to increase these, as they are particularly attractive.

JULY 5TH 1989

Oxford Mail reporter Lucy Tennyson came 2.30pm for a feature in her newspaper. *What a lovely feature, — thank you Lucy!*

July 6th 1989

Four Birmingham Roller pigeons, (now almost extinct, I'm told), have been given to me for the Sanctuary. We made a flight opening for them to fly out of the aviary after a few weeks of being confined. They now fly free, tumbling quite dramatically but keeping quite separate from the other pigeons and doves who fly freely around at the Sanctuary.

July 8th 1989

Purchased a pair of East Indies ducks from Mr. Boyce of Deddington. These small, rare, black ducks have a beetle-green sheen in the back feathers — very beautiful.

July 9th 1989

Sixty-five children from Cropredy School came and all thoroughly enjoyed visit; they found a baby rabbit in the warren outside.

July 11th 1989

A delightful coach-load of five-year-olds came from St. John's School, Banbury. Before leaving all were asked to pray for the success of the Sanctuary, along with their teachers. It was a sight I will never forget, seeing them standing among the hay bales with hands together offering up a prayer of thanksgiving to our Heavenly Father for the privilege of being able to come to Wigginton and enjoy the animals in the freedom and beauty that abounds here. Many visitors comment on the relaxed atmosphere of the Sanctuary as very few restrictions are placed on visitors. The safe surroundings allow parents to let children run round and be themselves — the reason why they all want to come back again and again. The Barn Nursery School came on the same day — they were delighted.

July 12th 1989

A nice letter of thanks from Tessa Lacey for my contribution to her book *"All about Geese"*.

Toddy Hamilton-Gould brought Daniel, a black-and-white British Alpine goat, one of twins born May 4th 1989, (castrated). He has a lovely mischievous nature, running circles around us, trying to trip us up with his tether.

Bucket Pond-filling

Very, very hot, no rain: two of the shallower ponds have dried up, – I had to keep buckets of water constantly filled. One haughty lady visitor commented *"of course you could be reported to the RSPCA"* as one bucket happened to be empty at the time of her visit. My response to that was rather heated, – *"Oh yes, you do just that madam, try to close us down and deny pleasure to all the thousands of others who love this place"*. Then the next group of visitors asked if they could carry water to the dry pond areas, and started individually to top up any emptying vessels and suggested that I put a row of buckets, filled with water, inviting visitors to top up the containers. This I did immediately and it proved very advantageous all round as many love to feel involved. This lady's remark was the only single hurtful thing that has been said about my whole Sanctuary since its foundation.

July 13th 1989

A Sunday School group from Enstone Church, on an annual outing, visited and signed the visitors' book (*"superb"*).

July 27th 1989

Some very good free publicity from the newspapers especially the *Banbury Guardian*. There is a nice picture of Samantha the rescued swan from Abingdon river.

Lucy Tennyson of the *Oxford Mail* devoted a whole page about my story. Had quite a few folk come in response and two ducks were adopted. Sold one of my little ducks to Southrop Pet Care, Hook Norton. She has gone to a good home.

My rabbits are now vaccinated against myxomatosis to protect them after seeing a wild one in the next field. Despite initial worried telephone calls to the local vet I was told the culture was not immediately available as there is only one supplier in the country. The matter was so urgent I called on an alternative vet who ordered the culture right away and came immediately on its arrival to do the all-important vaccinations. The rabbit warren here is top favourite with children and priority treatment for the rabbits was essential.

July 29th 1989

A nice few visitors today; a lady from Bladon brought children for a birthday party – with a birthday cake and food, balloons and squash. She adopted Samantha the swan and the children were very happy. First baby midget rabbits out of their nest.

July 30th 1989

Rain, rain, rain, — thank you, Lord! So many of our well-wishers have prayed along with me for badly needed rain at the Waterfowl Sanctuary. We have had no rainfall since we opened at Easter. The weather has been very sunny but do we need rain so desperately. Several more people came after reading the article in the *Oxford Mail* by Lucy Tennyson. She was more than just a reporter she was really genuinely very interested in it all.

"What's it for?" was the query of one executive type gent, with (obviously) no children. I replied that, if he had read the article in the *Oxford Mail* three days previously, his question would be answered. Since the Sanctuary opened on Good Friday, approximately 14,000 children have come, from tiny tots to teens. All love animals and feed the birds who come up to the wire asking for food. The goats, too, love to be petted and fed and make noises if visitors do not take a little pot of feed for them. Feed pots sell readily, most children come back over and over again.

The rabbit warren is the last exhibit to be finished and I got a gentle ticking-off from my bank manager, as I exceeded my agreed overdraft. If I had not had that warren made, a valuable piece of the Sanctuary would be missing. I am trying very hard now to put every penny into the bank, as my manager has been more than kind to me with this project. Interest rates are so high, it takes some doing to just pay off the interest, quite apart from the loan. Given nice weather through the school holidays, we pray dear Lord, that we get many more visitors. Some came to see Samantha the TV swan. She is so beautiful; I spent half an hour up at the top shelter, just looking down through the rain at Samantha and all of my other family. The ducks and geese love the rain but the hens and bantams are not so happy today, especially the ones with very feathery legs. We have only one baby peacock left at Wigginton out of seven hatchlings, very disappointing, plus two more young, hatched by a Maran hen at Enstone.

Part of my large and beautiful family, mostly descendants of Dad and Mum Taylor (above right)
Uncle Tom and Auntie Agnes (above left)

A

Dennis and myself at our wedding at Farnborough Church, on October 26th 1985 with ten little flower girls, relatives and friends

Our Sanctuary

Starting work with a JCB

Some of the first duck huts in place

Starting of the poultry pens

Start of the road construction

Poultry pens nearing completion

Poultry pens finished

Left: *Jackie, My eldest daughter,*

Right: *Son-in-law, Anthony, Lee on Penny the Pony, Micah in front and Stephanie*

Left: *Michael and Wendy, my second daughter in 1977 at baby Charlotte Rebbeca Page's Christening*

Right: *My son, Rodney Robbins, at the Christening in Wigginton*

Left: *Lee and Micah feeding lamb*

Right: *Granddaughter Melody with a near-hypnotised rabbit*

Ruthie and myself, 1994

Life-long friends, Joyce & Ted Soaffe "Hello my Duck"

Friends of the Sanctuary Brian & Pam from Milcombe

Joyce Watts, another dear friend, and myself

Fred and Marjorie friends from Kidlington

Kind Doreen Roberts

Volunteer helper Richard Valdambrini with Rhiannon

E

Left: *Alex Potter and friends*

Right: *Colin Turner, my friendly bank manager, with camera*

Left: *Edward Clarke, age 7, and his sister, Eleanor aged 10*

Right: *Paul Usmar, a kind young man in the early years*

Left: *Lorraine Angel with Neithrop School children*

Right: *My dear friends Pauline Sones and Gary Wilkinson*

Above:
*Hook Norton
Rural Fayre 1992*

Left:
*Melody
and Micah*

Right: *Friend
David Warner*

Left:
*Nephew Gary,
myself and
Andrea*

Right:
*Deron and
Brandon on
Thomas the
Tank Engine,
1997*

G

Two year old Nickie Harle from Banbury, 1989

Boys from Bloxham with Trixy the Goat

Murial from Wantage spinning, with Micah and Melody

Vanessa and Craig Lawrence and Alice Buckland from Buckingham

Siobhan and Mum, regular visitors

Penny proves very popular with Melody and friends

H

SAMANTHA THE SWAN

This is the story of a swan who staged a sit-down protest in the middle of a busy Abingdon road……because she didn't like deep water!

Along with many other swans Samantha lived on the River Thames at Abingdon, apparently serenely, until one eventful day something hurt her foot under water. The RSPCA could not decide whether it was a large pike fish which had given her a nasty bite or whether a propeller from one of the many boats had hit her. Whatever it was, Samantha was definitely unamused – and left the river. Her preferred place was in the middle of a busy road where cars, lorries, motorbikes and bicycles did not appear to disturb her.

Following very concerned telephone calls from members of the general public, Mr. Albert Honey of Animal Rescue removed Samantha from the roadway and took her back to the river. She returned to the road…… he took her back to the river. This process was repeated several times even though Mr. Honey thoughtfully provided a rubber dinghy for Samantha.

As a last resort, the Animal Rescue officer appealed on television regional news for a new home for the swan – preferably a dry Sanctuary. I saw the appeal, was captivated and offered Samantha a home. The RSPCA visited my (then) newly-opened Sanctuary and decided the surroundings were ideal for Samantha.

Mr. Honey brought her here in his van, Samantha placidly sitting in the back, wings secured to stop her damaging herself or anyone else. A swan's wings are so strong they can break human bones. Gentle hands carried Samantha uplands to the shallow pond where she immediately felt her feet on the bottom. Her acceptance of her new home was immediate and captivated viewers on that evening's *Central Television* news. She was cheered by several visitors, including pupils from Steeple Aston School, who promptly adopted her.

Samantha now feels perfectly safe in the shallow pond with a few other ducks and geese. She feeds gently from Rodney's hands taking her grain from him and talking quietly in her own swan language. Now that her limp has almost gone and the swelling disappeared, Samantha happily grazes the grass. Last spring she disappeared briefly for two outings to nearby fields, but was soon back. This time there was no road of busy traffic but just the lure of lush green grass in which she sat neck-high. She is admired by many visitors and has been photographed many times.

I tried to introduce the two young white Chinese geese into the pen with their own parents. They do not take kindly to them, pecking one and making it quite lame, but it now seems to be recovering. It is difficult as they tend to come off the nest with one baby, leaving all the other eggs to go cold in the shell. So, I took away the first two and she hatched another two so that's quite good, four lovely white Chinese geese, (I have yet to know their sex). They are the proudest looking geese of all, in my opinion. I only have two grey babies; they are nearly as big as the parents now. The mother, a White Chinese, instead of hissing and warning one off when sitting on her nest, flattens herself, faking death by lying quite motionless across the nest, allowing me to examine her eggs for fertility when necessary. It is quite a sight to observe as she makes herself as flat as a pancake. The other geese sit quite high. The grey Chinese built a nest quite fifteen inches high, laying eggs right to the bottom. I left them entirely alone this season but I shall incubate at least half next time. I'm risking a savage peck from my greys; they are totally different to the white. My one white was still trying to hatch a few eggs as late as end of July, but to no avail. What wonderful faith and patience those beautiful birds have! The female was bought from Chris Ashton in Wales. The male came from Shatterford Lakes, where he looked so very lonely all on his own. He mated up with my beautiful little lady straight away. They are inseparable and now number six in the family.

JULY 31ST 1989

We put our young runner ducks out on the open ponds, one white, three black, seven trout, one fawn. Approximately a hundred visitors yesterday, very good for a Monday. Cousin Frank Hawtin, Dennis's long-lost relative, called in from Australia – it was nice to meet him.

AUGUST 1ST 1989

One hundred and fifty visitors — quite good. The rabbit warren is still favourite.

AUGUST 2ND 1989

Lyn from Great Tew kindly brought in photocopied leaflets. Thank you, Lyn, and thanks to her boss for his willing co-operation. Around one hundred visitors today. A young lady came from London to adopt a goat and gave £10 instead of the normal £5. She reminded me of a certain star in *"The Good Life"*, I hope she comes again as I am sure she will want to see Snowy, her adoptive goat. Age Concern. Fourteen OAPs came from Kennington, (one in a wheel chair). They loved it, all of them. Went to see Mum in the evening. At nearly 88 years, she is wonderful, still knitting and crocheting.

AUGUST 3RD 1989

Mini-bus came from Bladon, approximately 125 visitors. My prayers are being answered: I'm managing to pay off some of the small bills, but still worrying about the big ones. Please Lord, let more people adopt a duck, goat, rabbit or swan and buy some of my young stock.

AUGUST 5TH 1989

Little Daniel came and bought two baby ducks, Muscovies, £2 each. A Soay ram purchased, two years old, husband for Zoë. When introduced they held a little conversation between themselves and seemed quite happy with each other's company. At £15 he was a bargain not to be missed, hopefully the start of our own flock of Soays. I fed him some ration laced with herbal stress tablets as I had done with Zoë. The effect on them is wonderfully calming. It is so much better if animals are not terrified of humans.

*Zeb & Zoë, Soay sheep:
in 1997 their offspring number 31.*

Samantha has befriended the Ruddy Shell duck, Daffy, who was rescued from a motorway. He refuses point blank to be shut up at night, so hopefully at the moment they should protect each other if Charlie Fox should venture near. Samantha, of course, has never been housed at night.

I sold the first of my young cockatiels for £14. Oh, they did peck me when I caught them! It was my first experience of handling them and they were upset at being caught, yet before I fell ill for three weeks they would climb on to my hands quite trusting and friendly.

AUGUST 6TH 1989

A pet shop man brought four large terrapins in a six foot long tank: filthy black water, gravel, stones etc. I set to and cleaned the beautiful tortoiseshell backs. Rod and I cleaned the tank out completely, putting contents temporarily into another tank whilst I got them thoroughly clean. My children had small ones when they were young, but I had not seen any as big as these here. When on honeymoon in St. Lucia there were some captive ones enclosed from the sea approximately three feet across, quite huge. Native children climb on their backs; people also sell the shells resembling shields, as ornaments.

AUGUST 7TH 1989

Last baby peacock disappeared, victim of stoat or rat. It is so disheartening and no use trying to rear baby stock at Wigginton outside. We will have to do most rearing at Enstone in future. The two hens are fine. Becky Tustain voluntarily came to lend a hand last evening, helping with water all round. Thank you, Becky.

Paid John Underwood £364 for the special cages he had made for me, not selling much out of them yet though. Hope they pay for themselves in the future.

AUGUST 8TH 1989

A lady from Lechlade had recognised my voice from television and then read the article in *Oxford Mail* which seems to have drawn a great deal of attention to the Sanctuary. Paid for badges, another red bill off my back. I'm getting them down now — thank you, Lord. Baby guinea pigs, brown, born four days ago. A visitor watched Samantha having a swim, also eight geese having a tussle amongst themselves as much as to say: *"this is my territory."* Samantha queens it over all of them. Sold baby rabbit for £5 out of the new pens. Most of the wire for the Sanctuary was purchased from John Underwood Farm Supplies, Banbury.

AUGUST 9TH 1989

Visitors saw a stoat or weasel in broad daylight, the one who has been taking my babies. What can one do about stoats?

AUGUST 10TH 1989

I sold ten small mixed chicks £10; two young Muscovy ducks. Visitors returning regularly now.

AUGUST 13TH 1989

A visitor arrived with two runner drakes and another visitor came from Oxford with a tiny Mallard duckling. A bantam who had been sitting and hatched one baby chick took to this wee Mallard, mothering it as its own, after I had tried to introduce it to at least three other mums, each with only one solitary baby chick. They are late hatchings, one Buff Cochin, one chick, one Sumatra Game chick and yet another Buff Cochin with one chick. One can see quite clearly why the rare breeds ARE the rare breeds as most of them are so infertile. So far this year I have two silver Sebrights, one of each, glad to say, two Poland, two' Black Pekins. Time has been very limited in the 1989 season. I hope for slightly better luck next year. Large trouser leg Cochins have produced one chick buff, one partridge, both cocks, I suspect. Disappointing.

George, the black buck rabbit brought here several weeks ago is calming down quite nicely. I am able to stroke his back now, approaching him only from behind the ears. If approached from his nose, he immediately attacks, snarling. At some time in his life he has obviously been quite badly frightened by something or someone. If only animals could talk we would then be able to understand; yet they do, of course, in so many ways. His behaviour shows that he has responded well following my treatment, just like the chipmunks and the Soay sheep. Two more rabbits arrived, both very bad-tempered as I've come to expect. The owner told me she has had them three years and they had given her no pleasure, as they were so unloving towards her. However, I shall work on them, herbally, as with other bad-tempered pets.

AUGUST 14TH 1989

Little Lucy, the blue bantam is mothering well the wee Mallard duckling from Oxford. Approximately sixty visitors today. Cloudy, with welcome rain.

AUGUST 15TH 1989

An *Oxford Mail* photographer came to take photos of the black rabbit, George, in my arms, snuggling up to my face. *(See front cover.)* He is lovely now, with complete confidence in me, — so relaxed and happy. I saw him carrying hay in his mouth; I wonder if he's a Georgina!

AUGUST 16TH 1989

I'm seeing families return more and more, many Sanctuary visitors. Have had printed *"Friends of the Sanctuary"* leaflets. Last few eggs, various hatching, the children absolutely love to hold day-old chicks and ducks. It is delightful for me to see the expressions on little faces. One little boy, named Ryan said *"I love you for letting me hold your baby chick,"* and another little girl said *"I love this little chick as much as my dad."* My own grandchildren love them all, too.

AUGUST 21ST 1989

Rod is on his travels visiting Wendy. Visitors now number 15,000. More baby guinea pigs born.

AUGUST 22ND 1989

Over one hundred visitors today. A dear lady, Mrs. Gray, sent a £25 donation cheque for the Sanctuary, with no address. I sent a letter of thanks and a free pass through her bank at Abingdon.

AUGUST 23RD 1989

The rare occasional child is not so nice. One threw a baby guinea pig down on to the ground. I happened to see him do it and reprimanded him. Instantly the lady in charge said he was from Spain, (as if to explain it). Later, after he had left, I found a dead chick, which had died at his cruel little hands, I suspect. All of our lovely little child visitors absolutely adore cuddling baby chicks and rabbits. My pets are very tame and trusting.

AUGUST 26TH 1989

Two beautiful white geese donated to us aged four months. Lady shed a few tears at the parting, she must have become very fond of them.

AUGUST 27TH 1989

Bank holiday Sunday, David DJ. *Radio Oxford* invited me to their studio in Oxford, Summertown. At this moment in time I would rather be at the Sanctuary.

Mange treatment necessary for *"Scinny"* pigs, the name one little girl called guinea pigs.

Thank you my duck!

AUGUST 29TH 1989

Bank holiday Monday, 3.30 pm interview with DJ David Freeman at *BBC Radio Oxford* once again a nice free plug for the Sanctuary. About 350 visitors without much advertising, very good.

Lyn came to give voluntary help, and cleaned out aviaries. Samantha Swan is finding her wings, I hope she doesn't go yet.

AUGUST 30TH 1989

Very sad. Sir Peter Scott died, 79 years: I did so want him to come and see my Sanctuary one day.

Tessa Lacey

AUGUST 31ST 1989

Tessa Lacey came to visit. She is a friend of Sir Peter Scott and his wife, Philippa. Tessa is writing a book about geese including her Maurice and Phoebe. She was thrilled to bits with every aspect of my Sanctuary. Her visit was in response to me answering an advertisement in the *British Waterfowl Association* magazine for interesting geese stories. She was fascinated to hear of Sebastapols' origins coming from Russia in the 1920s. They are still as beautiful as they were then. My young have feathered well, with even more beautiful feathers than their parents. Tessa worked at Slimbridge for a year. Along with celebrities interested in all aspects of wildlife conservation, she once had lunch there with the Queen.

SEPTEMBER 1ST 1989

Banbury Cake has a front page story about the rabbits I have calmed with the aid of herbs. It is natural food from my fields, plantain docks, dandelion, dead nettle, hawthorn, wild plum and sloe leaves all containing health-giving food. Human stress tablets from the chemist, lots of love, careful handling, time-consuming stroking and reassurance that humans are not going to hurt them has worked wonders. In particular, George responded beautifully to kindness and, following my influenza illness, I spent lots of time with him.

Old English Rabbit, Jealous, "Jelly", with babies.

The two spotted rabbits, does, brought to me in a nervous, vicious state attacked their owner at every approach made and came for my feet each time I entered the enclosure. I separated them, as they seemed jealous of each other. If I went to stroke one, the other would dash across and attack, hence the names Jealous and Jo. Jealous produced one beautiful little baby, exactly like herself. On finding a one-day-old baby out in the rabbit warren, cold and shivering, but its little heart beating strongly; I tried to introduce it to the other does, one at a time. Not one accepted it, so I entered the pen where spotted Jealous had produced her baby in a nice warm box in the corner. I stroked her then rubbed my hands in her bedding and removed a small portion of her fur with which she had made her nest and wrapped the little orphan in it placing it beside her own baby. Next morning I checked, and lo and behold she had taken to the second baby, obviously suckling it as her own. Four days have now passed. Fingers crossed!

September 2nd 1989

Approximately a hundred visitors today. Rod manned reception as I finished clearing out 36 Calthorpe Street. It is the end of an era. I've been trading as a florist in Banbury for ten years. Relieved to think I can now live a peaceful life with all my birds and animals at Wigginton. It is so beautiful and tranquil there and I have such interesting visitors mainly kind, gentle people, all with a love of birds and animals. How I love it all! Snowy goat came trotting into the Sanctuary reception, nosing into everything as usual. She had chewed through her tether, another new rope once again in tatters. Rod dashed into town to buy a new one, a large dog collar this time. How long will this one last? It looks strong enough at the present. Poppy, Thomas and Daniel are behaving quite well now. Daniel's donor, Toddy Hamilton-Gould, came to visit and was very pleased with his progress. She said Daniel was exactly the same size as his twin sister and looking just as happy and healthy. Rather nice to have a comparison as I am very new to goats, never having kept them before. Her little girl said *"I wish I could live here all my life."* What a wonderful advertisement from a small child's lips.

Still my white duck signs are allowed to stay. Although some have been 'beheaded' near the village, the message still shines through. I am so grateful to the council for not having them removed.

September 3rd 1989

Sunday: about three hundred visitors today bringing the total to approximately 17,000 since Easter. , Joanna, a little eleven-year-old visitor to Woodstock from London came at 11am with her adult friend, Margie, who was looking after her. They stayed all day, Joanna happily playing with baby chicks, bantams, rabbits, and guinea pigs. She fell in love with a baby bantam chick and was going back to London to ask her mother if she could take it home.

Another little girl brought two rabbits to re-home: a large lop-eared doe, and a Dutch black/white.

September 5th 1989

End of the summer holidays, few visitors now, time for maintenance work to begin in earnest. Ministry of Agriculture officer called about salmonella. Anyone selling eggs is required to test every eight weeks, or stop selling eggs for human consumption. I must breed as many bantam hens as I am able and try to persuade as many people as possible to keep their own for fresh eggs. I wonder?

September 6th 1989

Muscovy duck hatched another six babies in top pen, possibly last of the season; there's no more sitting to my knowledge. Sold three more Muscovies at £5 each. Still a sprinkling of visitors.

I sat in the evening with the sun setting red over the hill, watching my rabbits at play in the warren. A small pure white one popped out of a hole, – unseen before, followed by a black and white one, then a black one with two white front paws, they are so sweet. I do love baby rabbits as I love all baby things. The other white one being fostered by Jealous, ("Jelly" as I call her now), alongside her own, have their eyes open today and are quite tame. She allowed two little girls to hold her babies and they also stroked her. After being so vicious with her owner for three or more years it is nothing short of a miracle. Cleaned out the new cages in Reception. They are designed especially to allow little children to touch and get close to the baby stock.

September 7th 1989

A nice letter from Tony Baldry MP pledging his support for Rare Breeds exemption on the salmonella front. He is still waiting to find a duck warden for his home village pond, thus enabling the village to once again have the pleasure of ducks at Wroxton. Mr. Smith Ryland, Master of the Warwick Hunt, came and I showed him around the Sanctuary. I'm still sticking out: *"no hunting across my land"*. I have a standing invitation from him to visit his collection of ducks. I dipped all rabbits and guinea pigs with *Alugan* antiseptic against tick. It was dispensed by the vet and I put it in a forty gallon water tank, boiled up two kettles of hot water, added water at blood heat temperature and dunked them one by one up to their necks, avoiding their eyes. They dried very quickly on this lovely sunny day and looked much happier, no scratching now. I burnt all the old bedding, hopefully this has checked the problem in time.

SEPTEMBER 8TH 1989

An enquiry for Buff Orpington ducks, (only pale ones): I decided against selling my breeding stock as that would have been foolish. Hopefully next year, I shall need to incubate Buffs eggs, as demand exceeds supply. They are a lovely breed. One little visitor was taken to another duck farm on Thursday and told her granny that she still liked Wigginton best. What more of a compliment could I be paid from a tiny child? She has been five times already with granny and grandpa. Yesterday she brought her mummy for a surprise. Proudly, (she is no more than three), she showed her mother the way and bought a bowl of food. First visitors' book is now full up, must get a new one. Transferred all species out of Reception barn into pens outside, including George the black rabbit, who is now in the warren with all the other bunnies. Three more babies appeared; sold two guinea pig babies for £3.50 each.

SEPTEMBER 10TH 1989

Sunday — a young man and his wife brought four young Angora rabbits, the most adorable rabbits ever seen: white, fluffy bundles of fur with beautiful pink eyes. I'll be very reluctant to part with them, – the offspring, maybe. I have always longed to have Angoras and now at the age of 58 my wish comes true. Bless you, my dear Den, for letting me be me.

I sold a Sumatra cock bird £15, my baby from last year, (handsome bird). A visitor brought in a white female turkey. It was saved from the Christmas dinner table, hopefully to pal up with my blind-in-one-eye Stag, also saved from Christmas. Den has plenty of wheat for feeding the birds at the Sanctuary – that should keep us going all the winter.

SEPTEMBER 13TH 1989

A pigeon brought in injured; also a Mallard duck – the vet put a splint on its leg.

SEPTEMBER 14TH 1989

Visited by my bank manager. Quite pleased with results so far this year, but prices must increase a little next year to enable us to stay open. Three Silver Appleyards brought in free, two ducks, one drake, very nice. Small species seem to be most sought after, demand exceeding supply. Next year will hopefully see many more tiny ducks emerge.

SEPTEMBER 16TH 1989

Lovely gentle rain making everything look a little greener; the ponds are beginning to fill up again. A few brave folk came with wellies, umbrellas and mackintoshes, the children enjoyed it. One little boy slipped on a wet bridge and fell into the shallow ditch, a dry one, luckily. Must put a notice on wet bridges, to warn people to be wary, wire needed on surfaces. A cheque was brought to the Sanctuary for £25 subscription from Swan Foundry in Banbury to adopt a swan.

Another big black rabbit came with a £5 donation to help with his care. A young lady leaving home could not take him to her new flat. She said she would one day like to create a Sanctuary similar to mine. It is very satisfying to see God's creatures respond to love and care and watch their characters change in the relaxed environment created here at Wigginton Heath. A beautiful white Leghorn cockerel brought in, quite magnificent with a huge red comb and pure white feathers. He was very gentle until I put him alongside a Buff Orpington cockerel. They started to scrap through the wire divide, so I had to remove Buff for a day or so to let them settle. At first most cockerels feel anger towards another one placed close by, but I find they soon settle down when they know their rivals cannot get close to their wives in their own pen.

SEPTEMBER 19TH 1989

Russian hamsters donated, put into Maxi mouse town. I don't care for Russian hamsters as I found them to be excessively aggressive: I regret to say they promptly ate each other until only one was left! Sold one more young Muscovy duck £5. Only half dozen visitors today but it is a lovely day at Wigginton. What a joy it is to be here!

SEPTEMBER 20TH 1989

I collected two bags of stale bread from Mr. Malcolm's bakery in Banbury. This helps enormously to feed the many hungry mouths, ever open for more. All looking very happy now, the ponds are filling up again. The recent few days of rain helped, but it keeps visitors away unfortunately! I'm busy making rabbit hutches out of my old shop counters divided into four sections. They are ideal to house the rabbits and guinea pigs when the weather becomes too cold for them.

Another kind gentleman brought in a pile of wire cage fronts which, incidentally, fit the counters perfectly. Two holes were drilled in the side of each, slotted in, (two prongs), then a piece of timber was nailed for a shelter at the side, a staple added for a fastener. We put on a back and the job was done in no time! Thanks to the thoughtful gentleman for not throwing away something so useful to us. I could not possibly afford more rabbit hutches at this stage. Thanks also go to the visitors who bring windfall apples, stale bread etc. for the rabbits and birds.

No thanks to the six people who got into the Sanctuary without paying, then bragged about it in the local pubs. As we could not afford a turnstile when we first opened, we trusted everyone to pay their entry dues even if I was busy talking to other visitors. I remember the six (all mouth and deliberately taunting). They are the odd one per cent of all the lovely people we get here at Wigginton Heath who really care for animals and birds and appreciate all we do for them.

SEPTEMBER 23RD 1989

A school teacher at the Warriner School who lives at Whichford, told me she had picked up a lovely rabbit near Brailes. It had lived in her kitchen for the past ten days and she had become so fond of him that she was loathe to part with it. After looking around the Sanctuary she decided to leave him under our care and love. I asked her to call in to see him, but told her not too often, (just to look, I mean), as we need paying visitors desperately and not just donation visitors. I hope this does not sound too unkind, as kindness is what the Sanctuary is all about.

SEPTEMBER 24TH 1989

Lovely sunny day, two hundred visitors. Tessa recommended to a family in Swindon a home with me at Wigginton for the family pet, Mindy, a white goose. It was another example of them hating to part with her, (they were moving house), but as soon as they saw the Sanctuary they were a lot happier at leaving her in our care. Mindy was ten years old and had lived in their small back garden all that time. The two big men had tears in their eyes when they left. I put her in with a Brecon Gander and they settled down together very well, both being non breeders. Then later in the day I rescued thirteen hens, Warren type, all moulting and looking very sad. Two were promptly taken back to our land at Enstone – not a pretty sight while moulting. Put one of my Silver Cock pheasants out with peacocks in the netted run as the males had started to fight one another in the aviary. They are magnificent birds, now in full plumage.

September 25th 1989

Lady from Chipping Norton came to buy one of my white drakes and two of my little week old ducklings. She brought in her other one for us to clip the flight feathers and Den did this for her. The poor lady did not realise how much has to be taken off and was quite worried. Her duck won't fly away now as clipping on the wing, one only, stops them taking to the air, they can then only fly one-sided and for a very limited distance. They go around and around and this keeps them at home. The same applies to adult chickens, at least the flighty type.

September 26th 1989

Let out Birmingham roller pigeons from aviary, three of the five came back the same night to roost, hoping the other two would follow. Vet's surgery rang from Brackley. They had a strange bird brought in and thought it was a moorhen, but when Rod went to get it he established that it wasn't. He is busy identifying it now. This evening I went to pick up a web-footed bird from *The Bell* at Enstone. One of their visitors had picked it up in the car park: black and white and looking like a baby guillemot. How had it managed to arrive in the Midlands? It may have perched inside a long distance lorry. As it is only a fledgling, it eats rabbit food mixture and looks quite perky. Still only three of the roller pigeons returned to us including the beautiful pinky-buff coloured one, thankfully. One perched inside the Reception barn, up on the rafters all night and 'christened' my chair among other things.

Rod is dividing up runs for separating pairs of birds. We never have enough pens for breeding and need about another ten. Gates have to be made and more housing – I shall try a pair of ducks and a pair of chickens or bantams in each breeding pen, from January 1st onwards.

September 29th 1989

Mum over for the day – she is wonderful for her age but cannot walk far now.

Oliver, a little baby owl, arrived from Mr. Tustain. It is cute with its big eyes. At some time it lost a wing from an accident with farm machinery. Mr. Tustain has kept it alive but has now given it to the Sanctuary. Must now make a cage for it.

September 30th 1989

Archy, our Hook Norton policeman, called with his bird books to try to identify the strange bird found at *The Bell*, Enstone. At first he thought it was a razorbill, then after much deliberation he decided it was a fledgling young puffin. One wonders how it came to be at Enstone so far away from the coast, its natural habitat. However, it is devouring rabbit food mixture by the ounce, and loves it. I shall let it out with the peacocks in the netted run to swim in a pool of water.

October 1st 1989

A beautiful day. Joanna, from London came by coach to spend the day here, as last time she was here the little girl fell in love with a bantam. This time, Margie from Woodstock brought her and paid for a baby chick for Joanna to take home.

A gaggle of geese

Naughty Goats

Telephone call: could we take in two small pygmy goats from Carterton? They arrived and are very sweet – grey and white, one named Brock (as he looks like a badger) and the other, Whiffy. Put them into the house and pen on the lawn temporarily. A lot of interest in the puffin and owls. Little owl ate his first tiny mouse since he came to us. Finished his cage and little house for him, also a perch very close to the floor. He managed to fly up on top of his little box inside his cage – remarkable strength in his one wing, his eyes are so bright. Penny the puffin seems very quiet today, I do hope she is well. Rod and I are repairing the corner fence in Poppy's pen, ready to take the new adorable pygmy goats. Rod was carrying one in his arms with the other one following. They were almost there then the other one decided it would not go through the gate. Back and forth it trotted, leading us a merry dance until it eventually condescended to go in and join its companion. Both are wethers, (castrated males), we put them in with Poppy, our Golden Guernsey. She is very gentle and a great favourite with our visitors.

Pygmy Goat, Brock

Today she actually got out of her pen and trotted down to the middle of the chicken pens asking for food, then followed me back after the bucket, – no bother at all. After an introduction she decided she would bound after the very agile young goats. Around and around they chased. Poppy pursued them round the back of her house. The side of the fence comes around making an alley at the side through which they have to come. She actually tried to trap them as they emerged from the alley as though she was playing a game with them! Then, exhausted, she went into her house and flopped down, energy used up for yet another day, the first of many happy days with us. They had come to us as their owner's children were afraid of them. Their horns can hurt little children, even when they are playing.

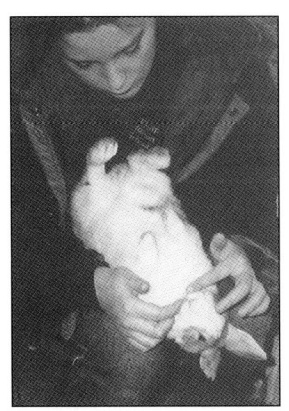

OCTOBER 3RD 1989

Baby puffin died, (sadly we were not able to give the correct food), and has joined the collection of other birds awaiting taxidermy in a deep freeze. Yet another rabbit brought to me for my own special therapy and herbal treatment. He is a lovely brown and white Dutch buck, who unfortunately had been castrated. The family had paid £20 to a vet to have him made unproductive. It is such a shame as he is such a beautiful colour. Unsure whether my therapy will work on him. Time will tell.

Pygmy goats settled down very happily with Poppy, showing her the respect she is entitled to in her established territory. Fun and games today catching our family of Soay sheep ready for dipping. They follow a yellow bucket now, but try to capture them, – oh dear! After many attempts we succeeded in penning them.

Melody holding a near-"hypnotised" rabbit

OCTOBER 4TH 1989

This afternoon we had to take the Soays to be dipped against sheep scab at our neighbour's farm. They broke away and around and around they chased, determined not to be caught. One by one they were secured again and all were dipped. Our little baby owl died, he was very badly injured in a farm combine all along one side of his body. His head was nearly bald, raw at one side. The poor little thing did not really stand much of a chance with those injuries. He was very thin, just feather and bones. It makes me so sad when a bird or animal dies.

October 5th 1989

I am making six more rabbit hutches out of old pine shelves. They are very strong and will last for years. Only one family came today, braving the wind and rain.

October 7th 1989

A party of youngsters visited today. Some had been before and asked if they could hold a baby duck. Unfortunately, we have none at present, having just sold off the last brood of Muscovies. There will be more in the future. I was able to let the children hold baby rabbits.

'Reaby', the Dutch castrated buck is responding to me beautifully. There is no biting or scratching at all so far. He lets me pick him up and stroke him, sits still and dozes off closing his eyes, loving being petted. One would never believe him to be the same frightened creature who first came to the Sanctuary. There is definitely an extremely calming sensation in the whole environment here. It appears to affect humans as well as the birds and animals. When visitors leave our gates they seem much more relaxed, shedding the uptight feelings of the pace of modern day living. Parents, grandparents and group leaders can bring children and, with a little supervision, just let them run free outside. There is nothing here to hurt them, apart from the odd tumble resulting in grazed knees, which can happen anywhere. Gentle nibbles of fingers poked through the goat pens has sometimes occurred, but these have not been serious. All ponds have been fenced completely, making it very safe for all.

October 8th 1989

One hundred and fifty visitors today, – a Sunday, lovely sunny calm day. Oh, I do love sunny days: such beautiful people come to visit the Sanctuary. Lynn, David and Claire, from Brailes, came to help water the goats and Claire picked up acorns from under the oak trees for the rabbits. Baby rabbits were cuddled today by many children and Mums and Dads. One man grumbled about the rough paths for wheelchairs – about the third person to do so since we opened. I pray someone will help finance tarmac for me, otherwise I will just have to wait until later. One dear old lady brought to visit in the late afternoon could not walk, so I invited her son and daughter to drive her around the perimeter road. Her son bought feathers for his Morris dancing hat. A young man, Peter, called. He sold me the young black turkeys from Middle Barton in 1988. They grew into magnificent birds.

Jean and Peter, an interesting couple, came from Standlake. They said they had been wanting to come since reading about me in the *Oxford Mail*. Jean rescues animals and birds, but is suffering from arthritis now and cannot bend down to collect eggs. It must be very hard for her, – such caring people.

I bought a pair of Diamond doves I saw advertised. They have the most endearing *"coo coo"* — music to my ears. I always remember that beautiful sound from years back when we used to visit Birdland at Bourton-on-the-Water. We bought a record with the bird sound on and that little voice, in particular, stays in my mind until this day.

October 10th 1989

Margie came and told me that the little bantam chick taken back to London by Joanna had completely changed the life of the little girl. It has delighted her very much, while before she was so unhappy. It's fantastic to think that a tiny bantam chick can make that much difference to a human being! God works in mysterious ways! Joanna let the little chick *"cheep, cheep"* on an answerphone message to Margie. I think that is lovely. Praise the Lord!

Geriatric Geese

I moved the goose, Mindy, in with the two other white geese but they pecked and quarrelled, so she had to come out. I tried her with the Sebastapol and after a little chat between themselves they were quite happy. Geriatric geese are not over-welcome, it would appear, with my younger stock, including Embden types. Brecon Buff are back together again and I am hoping to breed from them next year. My small ducks are separating into pairs now, one pair in each run along with the poultry. A stray wild cat is our latest enemy. It has caught and eaten one tiny call duck, the prettiest marked one. It was the one always outside of the wire. They puggle into the ground with their beaks until they make a hole big enough for them to squeeze under. Some are never content with whatever size pen is provided for them. Having sunk the wire into the ground for the hen runs we thought it unnecessary to do so around the sixteen ponds. Most of the ducks are quite happy to stay put, dabbling and puggling in the mud surrounds, but just a few explorers will not stay behind the wire, making themselves potential meals for the predators.

OCTOBER 13TH 1989

Separating pairs again. Four more geese arrived, all white, from a neighbouring farm. Margaret told me they kept pecking her sheep who lived in the same field. They would not leave them in peace and held on to the ears, not letting go – just hanging on, causing much pain. She was worried the rams would be unable to breed as the inside of the geese beaks contains sharp serrated gums, almost teeth, which combine with extremely strong jaws. If allowed to attack, these cause severe wounds on other animals or humans. Given herbs and our special treatment they will soon calm down. The main thing is not to let them see one is afraid of them. I did notice yesterday that Samantha, our swan, moved sedately away from the boss geese. When I fed them all they would not allow her to feed with them. She definitely gave way to them, although she is double in size and now eats in a different place to the geese.

OCTOBER 15TH 1989

A beautiful sunny day, approximately a hundred visitors again. One little girl, who has been visiting ever since we opened, brought a bunch of carrots. Her mother told me that she had grown them from seed all by herself, especially for the rabbits at the Sanctuary. She grew them between a bed of tiny baby field pansies. It is a delightful picture to conjure up in my mind. Her mum calls her *"Lockett"*. The family live at The Warren, just outside Epwell, a few miles from Wigginton Heath. Lucy is just three and a half and shows much love and care at such a tender age. The rabbits love carrots and most vegetables and fruit.

OCTOBER 16TH 1989

Three terrapins arrived this morning. Their donor, Mrs. Hall, is moving to Wales and this makes a total of seven at the Sanctuary now. Four of them are apparently twenty years old and the newest addition ten years old. These three had lived outside in a deep pond at Hook Norton for seven years

OCTOBER 29TH 1989

A very busy week with my daughter and grandchildren staying and I've had no time to write in my book. Now I must try to remember some of the happenings. Two lots of visitors returned to visit the rabbits previously donated by them to the Sanctuary, the vicious ones. Both were delighted and overjoyed at the change in their pets – no more grumbling and grousing as they used to do, deep down in the throat. Jeanette Gardener, from Clifton, came with her three children to bring a small wild rabbit. Among them they had tamed him down after he had been attacked by local cats on three occasions. He now responds to call and is happily installed with our large and ever-growing family in the outside warren.

Another small grey rabbit arrived in my absence, apparently abandoned by folk moving away. It was left in the garden and brought to us for a good home. He is a very friendly little chap.

Predator problems

Having a lot of fox trouble lately. Rats had earthed up the soil beneath one of the bridges and shorted out the electric fence. Charlie got beneath the fence and killed my beautiful male cackling Canada goose. Then the wind blew the electric fence apart in a tremendous gale and once again he got in and killed my baby white Chinese goose, this time the female, one of only four hatched and reared this year — so sad. I'm hoping now that I can acquire another mate for my cackling Canada. They are such a sedate gentle, unassuming miniature goose, one of the smallest type of geese in existence, not much larger than a big domestic duck and with a true cackle. I have not heard a sound from her since her mate died.

On another evening, in broad daylight, a fox came to one side of the Sanctuary while I was working on the other side and killed one Buff Orpington drake and attacked a Black Cauga duck. She was not quite dead and no visible wounds had been inflicted. I imagine she suffered a heart attack or something similar and she is now trying very hard to recover in the convalescent area. She is quite old, I believe. I hope she recovers. *(She was to die one month later)*. Then the fox killed my attractive mottled Muscovy drake, from whom I had been hoping to breed a new colour. He had recovered from a previous fox attack. That time he had a band of feathers taken from his body completely encircling him. Yet another sad loss for me. I wish so much I could afford security fencing, but at this time I am unable to and so I go on losing some of my precious birds.

Visitors Paul and Alison bought a rabbit from me. Her father made a hutch for her with a difference. He painted a mural inside it, with trees and bushes and blue sky so the rabbit would think he was still outside — (there's psychology for you). Alison's mother told me she kept an Aylesbury duck named Mary for ten years. She once laid a nest of twenty eggs, hidden away in an old barrel.

I'm trying to keep the six goats together in the top field; but no! – they find their way through, and back they come down into the Sanctuary reception area. Now we are having to put up sheep netting and I hope that will do the job. I'm thinking of giving the small pens a rest for the winter. Our family of six goats have other ideas, eating almost every tree and bush planted when they escape.

November 10th 1989

Not many visitors now as the weather worsens; it's not so pleasant in the open countryside these days. Our main concern is keeping our many mouths satisfied and happy with straw bedding and plenty of hay for rabbits and goats. Continuing the fencing in top fields to provide grazing safely contained in eleven acres for the winter months. It's quite sheltered from the wind in many places up there. The views from the nature trail are magnificent looking across as far as the eye can see towards Wigginton and Chipping Norton.

November 16th 1989

Telephone inquiry — would I take a family of rabbits? I agreed. Thirteen of them arrived, the main one a doe with four young ones, two of which were an unusual pretty mink colour. The mother had lost her ears in an unfortunate accident when she was a youngster and now, with her browny grey colour, resembles a large squirrel or a small kangaroo. Her former owner said she loves having babies, producing one litter after another, biting her way out of any cage to get together with a mate and even starting another family almost before her present one is weaned! The day after this consignment arrived we were clearing up the remaining hay bales, placing them all around the rabbit warren and, as we moved the bales, a baby rabbit dashed out and I caught it. Unbelievably, it was the same mink colour as the babies brought in yesterday! Most unusual! It must have been born to one of my escapees. There are a few who live out and around the Sanctuary quite happily. I introduced the baby to the others, but Nibbles, as she was called, rejected the little stranger. It was old enough to feed itself and was then put with other older babies. Then four more young rabbits were brought in, two of which were Old English black and white spotted, the same as Jo and Jelly, the vicious sisters. I hope for more of them to breed in the future. Introducing mixed blood streams results in strong healthy pets for many children to enjoy. I found a home for another big rabbit.

The tame mice are reproducing at an alarming rate. The old Maxi car placed for them to live in will be full to overflowing soon. There are black, white, grey, grey/white ones; it's fascinating to watch them busily going about their daily chores. They carry vast amounts of bedding from A to B and then back again, keeping themselves busy all day. In the evening they all come out to play.

NOVEMBER 17TH 1989

The goldfish pond is also becoming quite full of young fish. There seem to be dozens of them reproducing from the small amounts I introduced last year when the pond was first excavated. They are a joy to watch, very peaceful and relaxing. I added two small silver Koi last week – half price from the Crofts pet shop in George Street, Banbury, which is now closing down.

NOVEMBER 18TH 1989

Brown and white Tourist Board signs will cost £1,600 and I must go ahead with them as they will really put us on the map. Fourteen will be erected at all approach roads to the Sanctuary and are expected to be installed next February, a contribution to the tourist industry in my own dear Oxfordshire. The duck design will indicate mileage. Having lived on the edge of the Cotswolds at Lechlade for ten years, we visited many similar attractions nearby as our children were growing up. This made me decide to create this Waterfowl Sanctuary in Oxfordshire as there is nothing else quite like it in or around the Banbury area. I think I shall be paying off my bank overdraft for many years; I pray that God will give me the strength to do so! Final demand for £565 from Cherwell District Council. The rates for Wigginton seem excessive for what I have. (Yet another bank loan). Still, I battle on. What lovely weather we are having now — sunshine every day.

Zinny the Goose

NOVEMBER 19TH 1989

A young grey goose with pink feet and a pink bill, resembling a grey-lag but slightly larger, was brought to the Sanctuary by a pretty young lady from North Newington. She obviously loves the goose very much having hatched it from her own pair of farmyard geese. Moving on, she was unable to keep her pet, so the young one arrived with us. Talk about a shadow — she just follows any of us around everywhere! It is the same as having another child to mind; she eats when we eat, taking morsels of sandwich, cake, biscuits, crisps etc. from our hands or mouth very carefully. She would drink from our cups if we let her; she unties shoe-laces at every opportunity, tugging away until they are loosened; she nibbles on any string or electric leads so we have to make sure none is trailing. She visits all of the other runs and pens staying close behind my daughter when she is replenishing the many water bowls. She totally ignores all the other geese we keep at Wigginton and really believes she is a human being! She hops up into the caravan where we sit for our snacks at lunch time and is quite happy so long as she is near us. We have to shut her in a shed at night but she is not happy in there, much preferring to be with people. However tempting, she must not come indoors, as she has not been house-trained and leaves her calling cards around the place to be inadvertently stepped in. Visitors next year are going to love Zinny, (as she is called), as much as we do.

November 26th 1989

Sunny Sunday — a good trickle of visitors again today. Yet another large fat "Old English" rabbit was brought in; this now makes a total of seven; hopefully we'll breed many next spring. Fifteen pounds donation plus two hutches. Poor, poor rabbits! I feel so sorry for rabbits kept in small hutches all of their lives, but I suppose for thousands of people who want to keep them as pets it is the only way. Dear God, let me be here at the Sanctuary to give relief to as many as possible. Help me, Lord, to make enough money to feed them all for many years.

Another kind lady adopted a duck and rabbit for £10. One family came from Princes Risborough; they were recommended to come. Another young man came to buy a pair of my Sumatra Game birds. I have just two more to sell now until next year when I hope to breed some more. I must try to acquire more new blood stock, i.e. hens if possible, as my cockerel is a magnificent bird. I lost one of my Cochin buff cockerels — just simply died very suddenly.

November 27th 1989

More sad news. An Old English rabbit gave birth to six babies and promptly killed each one of them and also bit a little girl's finger. I brought the rabbit in from the warren and she will come home with me to have a lot of extra love and stroking. I was quite confident I had cured her bad nature but obviously she needs much more love and attention and herbal treatment.

December 26th 1989

Following an article published in the *Daily Telegraph* on places to visit, approximately ninety visitors came, some from as far away as London; all said they would come back again in good weather. One lady of 94 said it took her back to her Land Army days in the First World War. She could not see particularly well but the sounds and smells (scents) brought back very happy memories for her.

Our dear old friend Jack Badger came. He helped us when we were young and enthusiastic, with little money. One never forgets kindness like that as I find really wealthy people are reluctant to help others. This man was an exception. God bless him! His comment: *"Ah, you would have made it without my help Mabel."*; maybe or maybe not, one never knows.

December 29th 1989

Anthony and Jackie took me to visit Slimbridge – very enjoyable. My small collection is minute alongside, but it's ever-growing and so lovely here at Wigginton Heath. Another buck rabbit, black and white, needs a loving home. Someone will want him.

December 30th 1989

I must please my bank manager as he has such faith in me. The last thing he wishes to do is call in the loan after all the hard work that has gone into the Sanctuary.

A New Decade

January 1st 1990

After much thought and soul searching, it was decided to increase the entry fee to £2 for adults and £1 for children.

January 6th 1990

Wet, wet days – but there were a few brave visitors. It's very muddy – I wish I could afford more tarmac paths. We are busy repairing and damming up ponds to raise the level of the water in all sixteen ponds, they are all running well now, – a joy to see. It gives such pleasure to see all the ducks and geese enjoying deeper water, diving, up tails all.

January 12th 1990

First two eggs from white runner ducks: I've started to incubate them artificially.

January 13th 1990

A beautiful Barnevelda cockerel was brought in to accompany my lovely Barnevelda hen. A kind visitor, Gill, Jack Badger's daughter, noticed my lonely hen when visiting at Christmas and, as she had a spare cockerel, she decided to bring him for the Sanctuary. Very kind of her – he was most welcome.

January 14th 1990

Nice sunny day with approximately forty visitors. One sought advice on her own two ducks, but mentioned that she did not really want to keep too many birds. My reply, *"well, if you catch the "Buck Dug"* (meaning the *"Duck Bug"*), brought laughter all round. Ducks certainly grow on one!

I get frequent enquiries for just a few hens or bantams to keep in back gardens. The sheer joy for a little child of looking up the eggs and then having them for a meal far outweighs the chores of feeding and cleaning, especially if the hens are allowed a run outside on the earth. They enjoy scratching around, foraging for slugs, creepy-crawlies, worms etc., (their natural food), and their yolks are rich yellow instead of the pale insipid colour of the battery hen eggs. They also become quite tame and make adorable pets, if reared from chicks and handled quite often.

We had a few visitors but not enough in entrance money to pay for all the food my large family consumes. I managed to get stale bread from the local bakers, unsold from the shops, that helps, especially with the pigs. A few visitors bring along vegetables and stale bread. All is very welcome.

January 20th 1990

My small incubator is full of various eggs: duck and bantam and a few hens. We fixed up a creep for the young, with heat lamps, in the wooden shed. It is quite warm in there, so it should be ideal, providing we can keep rats at bay.

January 26th 1990

First little bantam chick hatched, silky white, almost like snow after the storm.

January 27th 1990

Severe gales have hit our countryside, said to be the worst for three hundred years. They took the roof off our shed, at the top of the three acres; blew down fences near the car park; took out five panels at Rectory Farm; and many branches of trees are down. Otherwise, we've been extremely lucky: (good construction all round). I've been stuck in bed with a flu virus and dizzy head: I do so hate being away from the Sanctuary. When I returned, the rabbit warren door had blown open, so several escaped, including the fat Old English. It took several hours to round them all up.

January 28th 1990

Visitor brought in a blue Muscovy drake with a damaged leg, also a little brown duck Rouen Cross. Two baby zebra finches in nest almost ready to fly.

One little visitor named our first little chick of the year Sid or Sidella, depending on whether it turns out to be a cock or a hen.

Winter Worries

JANUARY 29TH 1990

Banbury Guardian are putting out a plea for food for my hungry mouths. *Fine Lady Bakeries* very kindly sent seven bags of bread, mis-shaped and past sell-by date. I gave them an *"Adopt-a-Duck"* card. Sainsbury's gave vouchers worth £15 to buy loaves any time. They also offered videos on cooking for raffle prizes, but I would have to go to London to collect them. No way can I afford to do that. Really the necessity is food, stale bread etc. However, it gave me an idea to have a video made about the Sanctuary in the future, if I can ever afford it.

JANUARY 30TH 1990

Repaired netting on top of peacocks' pen. The gale had broken all ties along the west side. All repairs completed by lacing discarded electric fencing wire together, so making a good strong seam.

Trouble with magpies taking eggs, especially from the largest pond. Three separate nests inside the housing and not one egg in sight.

FEBRUARY 3RD 1990

I'm moving all feather-legged chickens and bantams to Enstone to keep them dry and happy. About twenty people came in answer to my appeal in the newspaper. *Radio Oxford* put out an appeal on my behalf resulting in approximately twenty-five bags of tail wheat and barley from Ted and Marilyn Ivings, Church Enstone; also tail wheat from our neighbours, the Cherry family; also crushed oats, goat and rabbit food, bread etc. arrived.

The "Banbury Guardian" made an appeal for food,...

... so did the "Banbury Cake"

One very small white rabbit produced five babies on Saturday morning and I put a heat lamp above them to keep them warm. I hope they survive. Two more lovely grey/white Dutch female rabbits came, much too fat, but not very vicious. I never cease to be thrilled when folk bring in rabbits of any colour, they are all so different. We lost our large dark Brahamas; they were all quite old. One hen only left now. Luckily the big Buff Cochin survived the storm, only because we put him under a heat lamp.

(Unfortunately, he died later.)

Blue Muscovy ducks will not stay in the two pens where they were placed for breeding: two pairs just climbed vertically up and over six foot high wire. Anyone who has ever kept Muscovies will know how strong they are; their claws are very sharp, resembling fingers. They can inflict a nasty wound, but in spite of all that they still remain one of my favourite species of duck;

They have such a strong will of their own, forever flying and climbing out. Luckily I keep very few drakes as I seem to have a good strain here throwing a high percentage of females.

The white Embden-type geese are pairing up, but not in the way I imagined. I put two pairs together on one large pond with a large expanse of grass, but two flew out to join others. One became friendly with newcomers brought in from a neighbour's farm, two have skew-whiff tails, the only way to describe them! They look quite funny when I am driving them into their house at night, one tail veers off to the right and the other veers off to the left. Obviously, I would not dream of incubating eggs with that malformation, but they seem very happy and healthy despite one of them dipping her beak in a tin of creosote we had left unattended for a few minutes. These three roam free and happy, there were no ill effects. Geese are tough, they certainly are! I must find good homes for some of the geese. Zinny has proven to be a remarkable character and she has fallen in love with Danny our small black goat. She stays close to his side every day, but refuses point blank to go into his house with him at night for safety. So, every evening, we have to allow her to come back up to the aviaries where we put her in the first instance when she arrived with us in the autumn of 1989. She totally ignores any of the many geese we have at the Sanctuary.

Zinny & Danny

FEBRUARY 6TH 1990

A sad day, following the frightening storm. One of my black swans was attacked, obviously finished off by a stoat. It's almost impossible to imagine such a large beautiful bird being killed by such a small creature. I suppose there are a few people around the world who actually like staoats; I am the exact opposite; they have killed more of my precious birds than dreaded foxes. They creep beneath the electric fence approximately eight inches off the ground. I wonder if anyone can help me ward them off. Is there such a thing as an electric wire on lengths of narrow matting say six to eight inches wide insulating the underside, preventing it earthing out? Would someone kindly research and experiment? (Possibly funds could be raised.) Please get in touch with me at Wigginton Heath.

Early "Mabel's Midgets"

FEBRUARY 8TH 1990

Five tiny baby white rabbits are doing well under a heat lamp with a good little mother, a new generation of miniatures! I realise that there is a great demand for good-natured, small-bodied pet rabbits – those that do not grow too big.

FEBRUARY 9TH 1990

First baby duckling incubated.

FEBRUARY 10TH 1990

I dipped six of the older resident rabbits in *Alugan* to cure them of ticks carried on straw. These nasty little mites embed themselves into the skin causing soreness and eventually death, if not treated. The drawback of rabbits getting ticks in the winter is that they have to be dried out with a hair drier after dipping, a very long and tedious operation. Two guinea pigs also required dipping. *(A powder form has since been developed and dipping is no longer necessary.)*

Joanna, the little girl from London, rang me today to ask *"Do you remember me? I took the little bantam chick back to live with me in London."* I replied *"You are unforgettable"*. I told her that her friend had described how the little chick had completely changed her life. How can one forget that? I gave her advice on feeding, but she has obviously been caring for it well since last summer. Amazing, — in the middle of London!

FEBRUARY 12TH 1990

Rodney repaired the covered netted run housing guinea fowl "Gleanies" from Africa. The birds were caught up and put inside, as I believe they are a member the vulture family. On rare occasions they attack other poultry if they see weakness or listlessness. I have witnessed three such instances. Gleanies are not among my favourite birds but they make ideal watch-dogs by making an extra harsh noise when disturbed. Their colouring is attractive with white/grey mottled and edged feathers.

FEBRUARY 13TH 1990

An appeal for food in the *Banbury Guardian* brought a wonderful response. They are printing a further story of thanks on my behalf this week. *Banbury Cake* sent a photographer to take a photo of Zinny goose and Danny goat who are in love, *(see opposite page)*, but neither would co-operate in posing, turning it into a *wild goose chase*!

Sold four of my laying hens. We have too many layers now, I'm cutting down to twenty-five, the maximum of laying birds anyone can keep without having them tested for salmonella, according to the new Edwina Currie law following the salmonella affair. The rare breeds I keep do not lay eggs for sale; only very few are kept to replace stock, if I am lucky. Someone somewhere has to breed replacers to take the place of the present replacers within the egg industry.

FEBRUARY 15TH 1990

Marjorie, Fred's wife, kindly obtained stale bread, cakes and buns etc. from a shop, for the birds, and also one bag of bread pudding. Quote from Marjorie: *"better not give the bread pudding to the ducks or they'll sink!"* Ha-haa!

FEBRUARY 19TH 1990

People are so kind and go out of their way to bring small amounts of food which would otherwise be thrown away, to help feed the birds and animals. Several visitors today, giving me the uplift I need in February. Half-term for schoolchildren this coming week. Managed to get all of my little flock of sheep and goats out to graze.

FEBRUARY 20TH 1990

Fencing in the two top fields is complete at last, no tethers or collars for the goats now and they can roam free. The farmyard geese are being transferred to former goat paddocks. Wendy loves the Sanctuary. She made tea for visitors as there was a storm raging with hailstones.

February 21st 1990

Stuart, a young visitor came as his tenth birthday treat. How good that makes me feel. A nice sprinkling of visitors today, a sunny day; several have been before. To think an ordinary person like me can provide pleasure for so many more human beings, little children holding baby rabbits and chicks. Five baby chicks hatched today.

February 22nd 1990

I'm finding things a little difficult as a grandmother looking after three grandchildren and, at the same time, trying to make a success of the Sanctuary. Half-term for schoolchildren and some sunshine brought a good number of visitors today:

First goose eggs laid by Zinny were sold. They went to London for decorating. A small boy is going to try to hatch two duck eggs in an airing cupboard, and, if he has no luck, his father is coming to buy two live ducklings. A visitor told me she bought a duckling here last year who has made a wonderful tame pet. Lucy, the Muscovy, follows the family around. Maybe, as time goes by, more people will realise what delightful pets ducks make.

"Dear Mrs. Rabbit Lady"

February 23rd 1990

I received a card from a pretty little girl who visited last year addressed to *"Dear Mrs. Rabbit Lady"*. How beautiful: it makes my heart sing with joy. I have realised that being able to give happiness and enjoyment to other people is one of the most precious gifts in life.

February 27th 1990

Storms, storms and more storms! Roof repairs and netting repairs daily. Thomas, the goat, again in trouble, tangled himself up with his tether. It is sad to have to tether him again. I will have to find him a new home as he is such a playful bunter, not good for visitors. He is a very attractive looking little chap, at a distance.

Eggs, eggs everywhere now. No luck with first Khaki Campbell ducklings, very weak on legs; interbred I fear. I must wait until later and introduce new bloodlines.

February 28th 1990

Storms again today. Still no visitors, I cannot expect any in such bad weather. Squirrel rabbit died yesterday for no reason at all, she must have been quite old. It is a slight relief for me as I could never answer the question honestly as to what had happened to her ears. Many visitors asked me and it was a bit embarrassing not knowing how to reply. Many more babies being born each day now; black white Dutch has a nest. Angora white doe still has her six, born on Wednesday.

March 1st 1990

My second black swan found dead on the pond. It had pined for its mate, didn't eat and then the snow and the bitterly cold winds finished it off. I think we are too exposed to wind at Wigginton for these Australian water birds. Experience will tell. Sebastopol geese have laid five eggs to date.

March 10th 1990

The *"breeding like rabbits"* saying is a fallacy as they certainly do not all breed well. Around twenty five per cent of the does in my care are just not interested in breeding. The barren ones are allowed outside in the open warren along with all the young bucks. They dig dozens of holes and tunnels in the earth, are extremely happy and very friendly to the children who go into the warren to play among them. It appears to me after a good deal of study that the reason for viciousness is insufficient exercise, living in small cages and frustration. I have found it takes around six weeks to three months to turn their characters from abnormal to normal. I have not had many failures to date. Love, patience, freedom and, above all, showing no fear of them is essential. I seem to have a telepathic communication with rabbits. Maybe I was a rabbit in my former life as I love them all so much. Every one is a different character and nothing pleases me more than when the new babies arrive, watching them grow day by day and cuddling them from two days onwards. When they first open their eyes the little visitors help by cuddling them too, creating a whole new generation of happy, healthy rabbit pets for children from the Sanctuary at Wigginton Heath.

Boy scouts and cubs from Helmdon, Northamptonshire, came to the Sanctuary to plant fifteen willow trees as part of their offer of help. They also raised £27.05 from fund-raising to help feed the birds and animals. Thank you all.

MARCH 11TH 1990

My daughter and son-in-law, Jackie and Anthony, were left in charge of the Sanctuary while I went to Wales with Wendy and Rodney.

MARCH 14TH 1990

Two teachers and fifteen seven-year-old pupils from Neithrop School, Banbury came on an educational visit, doing a project on coverings: rabbits' fur, chicks' feathers, egg shells, pony hair, sheep fleece, goat fibre, Angora and Cashmere. They had a hands-on session with rabbits, chicks, eggs etc.

MARCH 17TH 1990

A good few visitors today, Sunday: car park filled, a glorious day. One little visitor said *"I love this place"*. Baby rabbits were cuddled by many children.

One tiny black one emerged from its nest in the young peacocks' aviary. A visitor actually saw the peacock pounce on it, shake it by the scruff of its tiny neck, and toss it across the floor, poor little thing! I dashed across to its rescue and then promptly removed the remaining five babies and mother into the Reception barn for safety. The peacock had lived quite happily with the large white mother but was quite startled by this tiny black mite appearing from nowhere. It was unharmed, but it could easily have been killed.

A family today witnessed a wild stoat encircling a wild mouse weaving back and forth round and round, mesmerising, then catching it and killing it instantly. Stoats are my worst enemies, but this was a very rare sighting for a whole family visiting our Sanctuary in the heart of the countryside. They were very excited.

The terrapins have survived the mild winter in the pond. They are now fascinating visitors by appearing from the water and basking in the sun on the banks. Thousands of tadpoles in the goldfish pond are growing well. I found homes for three of the big male rabbits today. They went to a very kind family with grown-up boys.

MARCH 22ND 1990

Visit from my bank manager. He said I had all the right ideas and just needed good weather and good luck.

Kathy, from the pet shop adjacent to the main road in Banbury, asked me look after five wild-looking young rabbits at the Sanctuary, because they were terrified of the constant noise of traffic thundering by. When introduced to the others in the warren they were totally different little creatures, playing happily and eating well, relaxing in the sunshine and tranquillity of Wigginton Heath.

Dear sister May came again with cakes and refreshments to sell in the catering caravan.

Tesco bunnies

Saturday, four visitors arrived with a wicker basket containing three tiny baby wild bunnies. They had been rescued from a JCB earth-moving machine which had dug them up on the Tesco supermarket site in Banbury. At first it was thought they were rats but closer inspection revealed they were baby rabbits. The men took them home and fed them with milk every four hours, getting up in the night to do so. Their lives were saved when I introduced them into a mother rabbit's nest, a New Zealand white. She suckled them along with her own litter.

APRIL 12TH 1990

Much to my surprise the pair of huge, geriatric, lop-eared rabbits, whom I had thought were way past breeding age, today produced a beautiful nest of babies. Horace and Hilda had been living happily together as good companions for over a year; Horace was removed immediately, then "hey presto" — three weeks later there was yet another nest of wriggling fur. Amazing! I think she had been holding these youngsters inside herself until she decided to give birth, as he was nowhere near her after her first litter.

Horace

I found a loving home for another large unwanted rabbit.

APRIL 13TH 1990 GOOD FRIDAY

The Sanctuary celebrates its first birthday: free cups of tea for all visitors, approximately three hundred. Some returned who came on our first day of opening. Word of mouth advertising seems to be very effective. The Tourist Board signs are up on the main roads, fourteen altogether, some at small junctions. It is wonderful to have them up after only a year.

APRIL 14TH 1990

Very cold wind, rain and hail, yet quite a few visitors came today. Sixteen baby ducks hatched in incubator, white calls, white runners, and miniature Appleyards. Sue Ryan has been helping with posters advertising the Easter egg competition in the *Banbury Guardian*. The Lord is good to us.

Jenny, our neighbour from the top of the hill, brought back Jack the Jackdaw who had escaped

Thank you my duck!

from the aviary. He had been brought to us following an accident to his left wing, had recovered well and was able to fly again. He had been found at Adderbury, brought back, but as it was late at night he was left temporarily with Jenny.

APRIL 15TH 1990

Phyllis Sandle* visited. I had not seen her since I left Lechlade twelve years ago. It brought back happy memories of days in the Cotswolds and the ten years we lived there.

Sunday School children from Witney came in a mini-bus; they stayed most of the day and loved every minute of it. A coach load of sixty came from Adderbury: cups of tea and biscuits all round.

End of Easter holidays. Three mothers with seven children arrived at Reception saying *"We have just come to cuddle the rabbits, not to pay to come in."* This kind of thing is just not fair to paying visitors and we desperately need paying visitors to survive. Sorry, this is not something we can afford to do.

APRIL 16TH 1990

Seven guinea fowl and two doves came today as the donor's neighbour had complained bitterly about their noise and destruction to the gardens at Edge Hill. I bought a pair of cackling Canada geese from Chris Ashton, Welshpool; a pair of Australian Shelducks came from Brian Boning, Norfolk, also a pair of Carolinas; a pair of Golden Pheasants were donated.

Roy and Den working on the adventure playground – unconventional, many ropes. Quite different to the usual playground.

Easter Monday – approximately three hundred visitors: hail, snow, thunder, nevertheless everyone seemed to enjoy themselves. The odd person thought it was free to get in. I wonder how they suppose it could be created and everything fed and cared for without the need for money? Ninety-nine per cent of lovely people fully appreciate the work that goes on here rescuing and saving unwanted and unloved animals and birds. Another lop-eared bunny called Buttons arrived, a frustrated male: very snappy and unfriendly.

APRIL 17TH 1990

Small fledgling blackbird brought from Oxford by a caring person feeding him on worms and canned dog meat. A real survivor, this little bird!

Bought a pet lamb from Marcus Hughes. Wendy is bottle-feeding him on her lap like a baby. I try to record the feeling of joy and satisfaction when I see little children's faces light up with smiles of pleasure when they realise they are allowed to hold and cuddle my baby ducklings, chicks, rabbits, guinea pigs and mice.

Visitors comment time and time again how relaxed and happy everything looks here. My answer is that they are all surrounded by love from every direction. God has put healing in my hands which helps to calm the most vicious and bad-tempered of small animals, mainly rabbits.

MAY 5TH 1990

Thames and Chiltern Tourist Board representatives came to enrol me for a fee of £65. I can now advertise throughout the regions in their guide books, (when I can afford it). Mention of this was made on *Radio Oxford*.

MAY 6TH 1990

Four unwanted fantail pigeons donated.

MAY 8TH 1990

Today a racing pigeon stopped by for a rest and food then flew on its way.

*I heard later that Phyllis, this lovely young mother and wife, tragically died very suddenly.

Chuckie

MAY 13TH 1990

The yellow cockatiel, Chuckie, has been brought back. He escaped earlier from unfinished aviaries. He was given into my care in 1988 when I still had a florist's shop in Calthorpe Street, Banbury. When he first arrived with me, he was very chirpy despite having been mobbed by wild birds who had pecked all his feathers off, excepting his head cover. He had the usual crest and orange cheeks.

MAY 22ND 1990

Two blackbirds released back to nature, after feeding them until they could fly again. Three peacocks donated.

MAY 23RD 1990

JCB earth mover deepened two ponds in top field, piped overflow down to join the others, trickle of water as there has been no rain to speak of recently.

MAY 26TH 1990

Racing pigeon with a damaged right leg: rested too long in a village so was brought to me, kept in an aviary, still has limp.

Chuckie, brave little chap

MAY 27TH 1990

The little ring-neck dove has made good progress and can now fly again after many weeks: (released).

The three young peacocks are settling down quite well and are happy. The black and white pigeon with damage to its shoulder is healing well. Such a lovely day, so many caring people come to visit the Sanctuary.

The newly-dug ponds have now filled up.

MAY 28TH 1990

Oh what a happy day: everyone looking so relaxed and carefree. A trench is being dug to bury the wire around the extension to the already extremely popular rabbit warren, incorporating a large heap of soil for them in which to make further burrows.

Late this afternoon a visitor arrived with a young bird explaining it had appeared mysteriously on her son's lawn at Great Rollright. Was it a duck or a goose? Immediately, I identified it as a baby Canada gosling.

Roger and Jan from Hook Norton, organisers of the Rural Fayre this year, invited me to take part with a Pets' Corner. They told me a lovely story of a friend of theirs, an old lady, who kept a goose for forty-two years. I wonder if any of ours will live as long? Maybe they will. My son, Rodney, loves geese above all of the water birds.

Peter from Middle Barton brought Silver Dutch bantams in. The small girl who came with the peacocks recently told me the reason for them coming to me. The Manor House is a study centre and the students at the end of their sessions fill in a questionnaire and the most frequent comment is *"get rid of those bl..... peacocks."* Sadly, the voices of these beautiful birds do not match their looks.

June 6th 1990

Six East Indian ducklings incubated after removing them from outside where they would certainly be eaten by some predator, e.g. a stoat, weasel, rat, magpie or crow.

White call duck yielded ten eggs. Peacock has laid eight eggs so far. I am incubating them all now as predators have taken their toll previously. Buff Orpington hen, one baby chick only. Buff Orpington duck: raised four, sold two. Cheque for £60 bounced – such a smart looking businesslike gentleman, too! Carterton bank. I didn't ask him for his cheque card (silly me); it taught me a sad, bad lesson in human nature.

June 10th 1990

A very wet, cold week, no visitors. Then Sunday turned out to be a beautiful day. A visitor arrived with twenty pigeons, black, brown and white. I placed them temporarily into an aviary with peacocks and put up more perches for them.

June 14th 1990

I let out some fantail pigeons as I felt they were overcrowded with peacocks. Out they flew, around and around, then came back on to the roof, stayed with me, thankfully, and now this is home. A lady rang to say she has a Canada goose gosling, and brought it over, a companion for the one brought in recently.

June 15th 1990

Pigeon with damaged wing: I strapped it up – it will recover. A lorry driver had stopped to pick up a tiny bird which he brought in, a pretty Black Cap with bright orange chest and black cap. It had severe head injuries and died.

Each day we have to move mothers with babies to safety and comfort.

My teenage ducks are best left roaming loose in the field then rounded up at night. I have not lost one to predators this year by this method.

June 17th 1990

Happy, happy day – many visitors. Discussed salmonella with bird vet: he has no literature on the subject just now. I sold one Wellsummer laying hen. One peacock hatched strong baby, thankfully. All of the fantailed pigeons have stayed with us, music to my ears the *"coo-coo"* sound is so soothing.

Swallows trying to decide where to build a nest in our Reception area barn. A section of timber was taken out to allow freedom of flight to and fro. I hope they stay.

June 18th 1990

Rain all day, two visitors came last thing. One was a hearing impaired girl who thoroughly enjoyed her visit!

Aaron

Aaron brought in a little rabbit, which had been born with three legs. Aaron is a gentle, giant African man with a winning smile. He loves rabbits and small creatures. His parents have a collection in the wilds of Zambia, guinea fowl etc. He is married to an English girl and they have a gorgeous baby girl named Ebony.

Mrs. Hughes of High Wycombe came to collect back her two Aylesbury ducks and give a donation. She had missed them so much after she had left them here for two days. Such lovely people one meets through the Sanctuary!

June 22nd 1990

Wonderfully happy day. Coach load of fifty-two children and adults came from Coventry. Ecstatic comments about our place.

June 24th 1990

Harry, an Austin vintage car enthusiast, called in for a cup of tea. He promised to take photos for my book. Harry is Tudor Photography of Banbury. *(See back cover)*.

June 29th 1990

RSPCA officer Terry Winson brought a large terrapin to add to our happy family living in the pond. Temperature is up to 75°F today, so they have all surfaced, along with literally dozens of goldfish; dragonflies with blue, green, red, yellow and black hues dart and glisten above. The moorhens are building a nest at the edge. This pond is a sheer delight to me; I could easily spend hours just gazing at its contents. All the plants, pond flowers and trees are growing well.

Danny, the black goat, came and nibbled the leaves off two willow trees. He is a pain. He jumps right over the top of the wire.

A visitor brought a pair of black Pekin bantams as she has a neighbour worry over the crowing cockerel. She cried a few tears at leaving them but knew they would be happy among my family. I exchanged one Appenzella Spitzhauben hen for them, the start, I hope, of many more Pekins, as I already have two more baby chicks, and also two blues. Young yellowhammer brought in, and then released.

Fêtes

June 30th 1990

Put on a 'Pets Corner' display at Swalcliffe Fête and received a £20 donation for Sanctuary funds. We took Thomas, Larry, Cochin Big Ben and Betty, his wife, three baby ducks, a selection of baby chicks, white mice, the grey Angora mother with her four fluffy babies, (loved by all), and two pink-eyed albino rabbits. It is difficult to know which of our large family to take to a show, as all are so beautiful in their different ways. I want to go on living for many, many years enjoying my birds and animals. My gorgeous baby rabbits I love most of all when they are around three weeks old. They give me such joy to see little children cuddling them. Also the pleasure on the faces of adults endears them to me. Wendy is a great help every day, doing feeding and cleaning out. She is working voluntarily just now.

Anthony and Jackie continue to work hard in their florist shop, *Forget-me-not*, in George Street, Banbury.

July 3rd 1990

Today a beautiful kestrel was brought in by a man who had seen a bus run over it, straddling its body. It was concussed for a time but luckily no harm appears to have been done apart from a bruised wing and leg on the right side. It will recover and fly again. The right claw appears to be withdrawn as if it is in pain, otherwise it's quite bright. Fed it field mice, caught in traps in our medieval barn at Enstone. I believe mice are the staple diet of kestrels in the wild.

A man brought a lop-eared rabbit, called Snuggles, who seems never to have been snuggled after biting its owner over a year ago. I calmed it down nicely after half an hour of stroking and firm handling.

July 4th 1990

Wendy and I had day out at the Royal Show which made a lovely change. I left Rod in charge and came back feeling extremely pleased with my family of birds and animals, all healthy.

July 5th/6th 1990

Terry, RSPCA officer, rang to say he had a Mallard duck with an amputated foot just collected from the vet's. Could I accommodate it? Of course I agreed. He arrived, also bringing five rabbits, including two naughty Silver Netherland Dwarf.

July 8th 1990

Pets' Corner at Hook Norton Rural Fayre: Wendy helped. Very successful – we made a lot more people aware of our existence.

July 9th 1990

Fencing again: trying to keep goats in is almost impossible. Danny got out again and has had to be tethered. He jumps clean over the top of barbed wire. The others, Thomas, Snowy, Brock, Whiffy, Zeb and Poppy are behaving themselves and all looking very sleek and healthy. Little lamb Soay Fawn is growing well, also Wendy's pet lambs are beautiful: Larryadne with Harry Boy.!

Released the kestrel, he flew away quite strong again, we saved his life. I don't really care for birds of prey eating mice, but that's nature, I suppose. The law is against feeding live creatures to other animals in public view nowadays.

July 10th 1990

A few visitors; the curator of Woodstock Museum called for leaflets. A little visitor from Sussex, staying with her aunt in Banbury visited yesterday and fell in love with a little grey rabbit. She held it for a long time then went round the Sanctuary but was soon back for more cuddles. She just had to ring and ask her mum if she could have it. Mum agreed, hence her return today. A rabbit is undemanding compared to human relationships where one partner commits his/her life to another. At least a rabbit eventually passes not too long away from this life after enjoying many hours of love and cuddles. The little visitor has taken the smoke blue Angora baby to east Sussex, so concluding a very happy holiday in Banbury.

Motorway Destruction at Fencott

July 13th 1990

An Irish family visited and enjoyed their visit immensely staying for four hours and had a picnic. The Father was very impressed with the natural hedges we have around the Sanctuary, all the original hedgerow shrubs of years gone by. At his home in Ireland the hedges are no longer the same as he remembered them as a boy, thirty plus years ago. They had been sprayed and cut back by machinery for modern-day farming. My land must be almost unique here at Wigginton in this beautiful part of the British Isles. Even Fencott fields as I recall when a child, are at this moment disappearing under the bulldozers for the construction of the London to Birmingham M40 motorway. Those lanes, almost grown over with dog roses, honeysuckle, wild plum etc, – how delicious were those hedgerow fruits, if we could reach them in our Cholsley field hedge, before the travelling gypsies picked them. It was a race to see who got there first!

Then there were the old apple trees gnarled and rubbed away by our cows who used them for shade in the heat of the summer. Every spring, masses of sweetly-scented pheasant-eye narcissi would push up through the hard earth in spite of many animal feet tramping the soil. Blackberries abounded in those days, their sale my only source of pocket money (apart from selling a few pet rabbits to friends for sixpence each). I used to gather blackberries each autumn and sell them to a man who came around our villages buying them for a dye factory, or so we believed. We neither knew nor cared so long as he handed us a few pence per pound for our labours. He came with a flat bed lorry carrying rows of shiny barrels. — After weighing our containers on his big scales he tipped our fruits out then weighed our baskets deducting the price of the container. Every year I looked forward to blackberry time to earn pocket money as Dad only sometimes gave us half a penny a week for sweets, gob-stoppers or sherbet dab with liquorice. What blissfully happy days they were at the old Homestead Farm at Fencott, (near Islip, Oxfordshire).

Runaway Combine

JULY 14TH 1990

I took a selection for a Pets Corner at Milcombe Fête: very hot, no shade, unsuitable for pets. Wendy took home some animals in my van at 5pm then returned to say Rod could not get out with the tractor and trailer as the road was blocked by a combine harvester that had lost control and smashed into a mini-bus full of people who had just visited our Sanctuary. They had been waiting in a lay-by, flagged down by the escort, to wait until the machine had passed by but the huge combine's brakes failed. What a terrible experience for the people: four adults and four children, no one killed but all taken to hospital.

JULY 19TH 1990

Next day the owner of the mini-bus involved in the crash called to say all were released from hospital the same evening, except one small boy with a cut face. Thankfully, no one was badly hurt.

Wendy and I went to Longhope, Gloucestershire to buy two baby kids (Angora goats). They have adorable little faces and are very cuddly and soft. We are calling them Roger and Bobby. They cost £20 each, which is very reasonable for little wethers, castrated males. Mrs. Alexander and Mrs. King run a stud farm of five hundred Angoras, also horses. Wendy's little kid, Roger, has a stomach upset, Bobby is not affected.

JULY 20TH 1990

A little owl was brought in, very sleepy; I put him in the incubator but unfortunately he died. I rang a taxidermist, who came at 5 pm and took the little owl, also a small pheasant that had been brought in recently, picked up on the road from Devon. These birds must suffer terrible trauma on the roads before they actually die. I do what I can, I strapped up the broken leg. He followed me over to Enstone to look into my deep freeze with its collection of frozen birds. He took my barn owl, also my little humming bird and promised to preserve the puffin and my little Pekin black hen in exchange for these two rare birds. He gave me a quote for preserving black swans, (a possibility in the future).

JULY 22ND 1990

Small pheasant poult brought in today. Many visitors: more people are finding us thanks to the signs on the roads.

My large partridge Cochin was mysteriously found dead in the pen – it must have been heatstroke as he was fine yesterday, – I'm very sad at his loss. He was so magnificent; hopefully one of his offspring will grow to be as magnificent as he.

JULY 23RD 1990

Hook Norton "Jack in a Box" Nursery School came: thoroughly enjoyed themselves. A very successful day, first day of school holidays. Three grandchildren to look after for seven weeks.

JULY 26TH 1990

Visitor brought twenty mixed bantams. Vet had to put the little honey-coloured rabbit to sleep after stitching its wounds. These were inflicted by a much larger rabbit when it was irresponsibly put back into the wrong pen. Sadly it had not responded to the treatment of its dreadful wounds beneath its body, poor little thing. I was sorry to say goodbye to it.

I bought a machine to clean my floors for £300 – a lot of money, but it's just what I have wanted all my working life, a really strongly powered vacuum cleaner that sucks up all the hay, straw, and grain etc. It is smashing!

JULY 27TH 1990

Lots of visitors again. Wendy brought a really fat brown rabbit from an old gentleman. It was very bad-tempered and tensed up. I sat and cuddled it for half an hour or so, clipped her nails, which had grown very long and uncomfortable, and soon she was relaxed and peaceful, the fear fading from her eyes. Two more, almost identical, not quite so old or bad-tempered, were bought in. They are happy now outside in the grass pens. Rabbits are taking over my life at the Sanctuary. I am organising a programme of babies to enable little visitors and big ones to come to cuddle them. Kids of all ages love to cuddle baby rabbits, the joy on so many faces gives me so much pleasure knowing other people love rabbits as much as me. Crazy, I know I am!

I encouraged a young lad called David from Sibford to get his pen ready for the ducks he so badly wants to keep by giving him a slatted construction Rodney had made, but never used. I find helping youngsters most rewarding, (not money-wise). David had described to me how he had made his pen so far, it sounded as though he had put a a lot of work into it already. Grandpa and granny came to collect it for him, and brought an interesting brother along, too, a horticulturist in the making. It reminded me of my youth and the early part of my marriage to Bill, the garden boy who gave up an academic career and left grammar school to become a nurseryman for the remainder of his life. We always kept ducks, hens etc. and worked hard at everything we set ourselves to do.

JULY 28TH 1990

Four small guinea pigs arrived, also a pair of golden pheasants. I put them in the pen with the others; the male was a very poor, weak specimen, and my silver cock-bird must have attacked it, as a visitor noticed it cowering in the corner. Nature is very cruel, the most dominant will kill the weaklings off; the female bird was left alone. I must find another male.

200 visitors today, including a group from Kidlington. I discovered I was actually at school with some of them. It was nice talking with old pupils, bringing back memories.

Duckling from Liverpool

Visitors from Liverpool brought a duckling found at one day old. They thought it was dead, revived it, fed and cared for it until it was three weeks old; then they brought it on holiday with them in their caravan to Aston, the other side of Stratford-upon-Avon. They heard about my Rescue Centre, came to the Sanctuary, and were thrilled with the place and with the fact they could leave their little duck safely with me. It was a nice companion for the little black one which hatched out quite recently.

Banbury Cake came to photograph Bobby and Roger, our Angora goats, with Melody, my granddaughter.

Melody with Bobby

AUGUST 5TH 1990

Visitor came with three baby field mice found on an allotment. I added them to the white, tame mother mouse's nest. Little visitors enjoy picking up little mice but their mums often say they give them the shudders. An amazing reaction to such tiny creatures!

A Welsh Harlequin duck was brought in with no feathers left on her back or neck – the same old story, attacked by too many drakes. It's happy now within our Sanctuary.

Twenty-six young pheasants of various colours were brought in by a man from Northamptonshire. Stoats have been taking toll of them. I'll have to move the babies.

AUGUST 6TH 1990

Banbury Camera Club came this afternoon. They spotted a stoat lurking nearby.

Kidlington brownies, staying at the Hook Norton Centre, visited.

AUGUST 7TH 1990

A family from London staying at Deddington came to visit.

AUGUST 8TH 1990

London visitors returned and took four white mice back home with them for 50p each. They just had to stop by on the way home, they loved my Sanctuary so much.

The most remarkable thing happened last evening when shutting up. I caught the sound of a baby chick coming from the Blue Pekin run and went to investigate. I saw a tiny chick about one or two days old. As no mother was broody in that pen, the only explanation was that the pigeon who had shared the hut laying her eggs must have incubated this egg as her own. The little baby tilted its head back and opened its beak so I assume the pigeon had been feeding it for a full day since it hatched, the same way they feed their own young. Then I removed it to the heat lamp, and placed it alongside the other day old chicks, trying to teach it to peck up the correct way, (downhill not uphill.) We are calling this little Blue Pekin bantam, Lucky.

Penny, Mr. Oldham's Shetland Pony

Penny

For several years I had noticed a Shetland pony grazing on the side of the main road at South Newington, and it has recently become law not to allow ponies to graze on the roadsides. Mr. Oldham had decided to retire and sell his little pony, Penny. He rang me to ask if I would take his budgies into the Sanctuary and I agreed. Then I asked what was happening to his pony. He said a horse-dealer had offered £100 for her. I immediately offered £110 for her if he brought her along tonight. He happily agreed and she is now grazing safely in our paddock. Rod made a horse shelter for her from off-cuts — but being a Shetland she prefers rain and weather on her back and seldom uses it.

Water still hardly running, we have to top up water vessels with mains every day now, but the glorious weather keeps the visitors coming.

AUGUST 10TH 1990

No mains water in taps now at Wigginton or Hook Norton.

August 11th 1990…

Still no water, I'm having to bring it in tanks from Enstone for the birds and animals.

August 12th 1990

The ponds are holding just a little water in the bottom. Please, Lord, let it rain soon.

August 14th 1990

Still no rain to speak of, all our hay is in the barn.

August 16th 1990

A moorhen, brought in by Nick Allen, almost dead – died one hour later. A fishing line from Adderbury Lakes was embedded into the knee and ankle joints. The poor little bird must have suffered a very painful death, as it was quite unable to walk, and could only flap its wings to drag itself along the ground.

One blue Pekin bantam died, sad, as I have only reared two blues and two blacks to date. So far this year I have also reared eight Buff Orpingtons, seven Partridge Cochins, three Appenzella Spitzhauben, five Andalusian, one Sumatra, one Gold Sebright, sixteen Rhode Islands, one Light Sussex, seven white Silkies and one blue Silkie.

Bank Holiday Monday 1990

Took a Pets Corner to Hethe Victorian Fayre. This proved very popular with the children once again. Sold one rabbit and one mouse and £20 was paid for attendance money – not really worth the trouble of packing everything up and leaving the Sanctuary for the day. Rod stayed and manned the counter, assisted by kind Sue Ryan, giving her support, just as so many well-wishers like to do, for the good of the Sanctuary. Some people are so very kind.

August 29th 1990

A group of disabled people from Chipping Norton visited — all enjoyed it immensely. School holidays nearly over. What a success story: every day many visitors, and still no rain. Ring-necked dove brought in, unable to fly.

Peacocks

August 31st 1990

One aged lady visitor came marching in thinking she was at a Slimbridge-type reserve, announcing that she didn't like to see birds caged in any way. She was commenting on one pair of young peacocks housed in a twelve foot square cage, which is quite adequate for birds of this age, as they have not yet grown their long tail feathers. I told her that if she would like to leave a substantial donation I would immediately have a very large flight pen made. Young peacocks cannot be allowed to wander free as they fly away into the countryside and get eaten by foxes and mobbed by wild birds. One of ours escaped and flew towards Hook Norton. It had rained at that time and we had to go into a field of waist-high grass and get soaked. With the help of the kind farmer, who spotted the bird, we retrieved him, and promptly put him into the large pen, measuring fifty feet long by seven feet wide. The sight of peacocks flying free is for many people, including me, one of the most beautiful scenes that can be conjured up. But we are set in hunting country, not safely inside castle walls, twenty feet high, (the ideal environment for such birds). Just now my resources don't stretch far enough to do everything in absolutely ideal conditions. Thousands of other visitors, however, just love looking at all my birds, never complaining, just thinking how very beautiful they are.

A Netherland dwarf rabbit was brought to me from London for my therapy for a week's treatment. The little girl could not handle her at all but after I had demonstrated how to hold her, she walked slowly away carrying her beloved pet.

September 1st 1990

Two more animals have been brought in for therapy recently. One is a Giant Himalayan rabbit which had been frightening the whole family. He would attack feet and legs, the man having to wear the protection of wellington boots when he went into the garden. The animal would bite, scratch, grind his teeth and grunt. Now, one week later, he is quite a different personality. The second rabbit, an old English, has also calmed down, helped by my healing hands and massage; the tension has gone from his body and the family are very pleased and have taken him home.

Emma from Nuneaton brought a drake named Thomas. This little girl had the little duckling since it was tiny. It grew to be a very large Muscovy and was literally imprinted upon this little lady, following her everywhere she went. Emma, with her uncle Brian, came in first to see what kind of a home they could expect for her pet duck, whilst her sister stayed out in the car park holding Thomas. Emma OK'd the Sanctuary and having seen my huge happy family here, she went out to fetch her adored pet. Tears overflowed at the thought of leaving him. It was a scene I often witnessed in the florist trade when dealing with bereaved people. Emma took her pet and placed him in one of my pens, together with a young duck, and he started his courting straight away. She said she knew he was going to be happy here, as he always followed *her* everywhere, but he was now busy giving his attention to his new-found "girlfriend." The farmer next door to her home had forced her to part with her pet as it kept flying into his yard and perching on the roof of his new car. Anyone who is familiar with the claws of Muscovies will be slightly sympathetic towards him. Emma left feeling much happier and hopes to comes back and visit him from time to time.

Thomas & his "mum", Emma

September 3rd 1990

Two Aylesbury type ducks and three runners came today.

September 4th 1990

Aaron, the kind gentleman from Africa, called again, bringing a golden guinea pig and a small black rabbit. He regularly rescues unwanted animals and birds. He reminds me of his junior counterpart, Daniel, who, when I first knew him, barely reached eye level with the Reception counter. He asked me, *"Have you any Coscoroba swans?"* Taken aback, I had to admit that I hadn't. Then he went on, *"Have you any Trumpeter swans?"* *"No, my duck, I haven't any of those either."* This charming little black boy seemed to know all the names of the different birds and I'm told he always wants to read books exclusively on birds. In fact I only had Samantha, the TV swan. Daniel's only ambition is to be with birds and I hope and pray he can find his rightful place and home in the world with them. He lives with his foster-mum, Diana, and also Jonathan.

September 5th 1990

A tawny owl brought in. I tried to feed him with mice, his natural food, but he refused to eat, so he was taken off to the vet's at West Bar for observation. A really beautiful bird.

September 6th 1990

Went to Gloucestershire today to buy three lavender and three buff Pekin bantams, as mine were disappointing this year.

September 8th 1990

Four smallish terrapins were brought from Long Crendon to add to our other ten already established in the goldfish pond. They live very happily on the weed and natural plant life and come to call when fed with pellets, as do the goldfish – very entertaining for visitors. When the sun shines, terrapins venture on to the bank and bask, enjoying the warmth on their shells or they lie on the top of the carpet weed. Today I spent time dragging some weed from the pond.

September 10th 1990

A lady rang this evening to say her husband had found a badly injured piglet on the side of the road between Bicester and Deddington yesterday evening. He had taken it home and fed it puppy milk. The RSPCA had advised them to bring it to us. I took it to the vet, whose opinion was that it was so dehydrated it would not last the night and was barely alive. However, he injected it with antibiotics and showed me how to force feed the small pink baby, feeding special piglet colostrum by inserting a tube down into its stomach. The back legs were broken and the little hips crushed with severe bruising to the side of the head.

September 10th 1990

The piglet brought in yesterday sadly died, despite keeping it warm through the night with a light bulb, electric blanket and hot water bottle The vet thought the only explanation of it being by the roadside was that a fox had taken it away from its mother who had her babies in a pig ark outside in a field, and had dropped it crossing the road on its way to feed its own young. How cruel nature can be at times!

September 13th 1990

One of my beautiful white fantail pigeons was killed last night, eaten by a stoat or a rat. It was not flying as others do, but just sat and walked around on the ground, nesting beneath one of the hen houses, out of my reach. On the other hand, many more have hatched and fly happily around this year, I am very pleased to say.

A black pigeon found its way into a water butt and drowned; it was just floating like a water lily, sleeping peacefully on the water. The lid had been left off the tank for just a little while and the bird could not get back out up the smooth sides. All birds need to have ponds, and ideally drinking vessels, with gently sloping sides. It is sad, indeed, to lose one in this way.

September 14th 1990

The extension of the rabbit warren is now completed giving them one eighth of an acre in which to bury and make burrows in the ground. It will give the visitors more space to go in among them, as the other warren used to get quite crowded on Sunday afternoons. With the maternity block enclosed against the weather, all is looking very comfortable.

A pig-sty is being built by Roy with oak posts and thick strong off-cut cladding. We have been promised a Vietnamese pot-bellied pig as a family moving house could not take this unusual pet with them. Our vet arranged for it to come to the Sanctuary. It will have a large run outside to live in and root around, quite close to the adventure playground and next door to the rabbit warren. This type of pig is very friendly and enjoys the company of human beings and can be house-trained like a dog. It loves to live indoors and the latest trend is to own a pot-bellied pig and acquire a pig sitter instead of a doggy sitter when off on holidays or outings.

Another town pigeon with a damaged wing was brought to us via the police, from the landlord of the *Three Pigeons* Pub in Banbury, who kindly took back leaflets and a collecting box.

September 18th 1990

I collected a tawny owl from the vets. Although blind in one eye, it looks much recovered after two weeks care in their hospital. He sits on a perch well and tries to fly, spreading his wings, but seems not to want to eat. We have to open his very strong beak and push part of a defrosted day-old chick down his gullet. The wise old eyes are only half as wise now; he allows me to pick him up and seems to be gaining a little weight. When he came in to me he had absolutely nothing in his stomach and could easily have died of starvation. He appears to be unable to pick up feed for himself, so three times daily we force feed him, one of us opening the beak and inserting the mouse or chick. (Owl food is supplied to zoos and bird parks in frozen form, then must be allowed to defrost before devouring.)

He was found by the roadside and had apparently been hit by a moving vehicle causing temporary concussion and then starvation. Just one kind person stopped and turned his car around and went back to pick up this beautiful bird in an attempt to save his life. Ideally, I would release him back to the wild around the Sanctuary where other species of owl live naturally, feeding from our unsprayed land. Until he can feed himself and fly well, however, he is better protected in a cage, albeit a very large one.

September 21st 1990

Third day with us and tawny owl managed to devour a whole mouse in one gulp. Luckily, we have plenty of wild mice here. He shows great improvement every day.

I am a grandmother again. Jackie has a little girl, she is beautiful, just like her brother, Lee.

September 23rd 1990

The tawny owl, now named Ollie, is progressing well, regularly eating without having to be force fed. I'm very pleased. Yet another pigeon brought in to us with a damaged leg, but otherwise it's fine. It will soon be flying around the Sanctuary with the ten or so different kinds we have living freely here. Several insist on coming into the Reception area to perch on the roof beams. They make a lot of mess and I would prefer them to stay outside, but they are attractive.

I went to a farm sale: bought lots of feeders, drinkers and troughs, — very cheaply.

Emma's Frizzle Bantams

September 24th 1990

Three Frizzle bantams were brought from Uxbridge by a lady who had visited our Sanctuary back in the summer. She could not bring herself to kill off these two cocks and one oldish hen. What pretty little bantams they are, very tame too, trotting happily around the barn area.

We are trying to persuade the pigeons not to use the Reception barn for roosting as they make such a mess every day.

September 25th 1990

George brought blue Cochins, which are large birds. I gave him some finches, although I dislike catching the birds in the aviaries, it upsets them so much.

September 26th 1990

Purchased eight yellow canaries from Charlton; also Cocky, a Ring-neck Parakeet, very noisy and quite nervous at the moment; he is a lovely shade of green with a red ring round his neck and he needs a wife.

Cocky & Ruby close companions for six years

September 28th 1990

I bought a trap to catch wild mice for Ollie owl, alternative food to frozen day-old chicks. He is progressing very well indeed, still blind in one eye. He flew out of his cage and landed on the ground, but allowed me to pick him up and put him back.

September 29th 1990

Today I received another lop-eared rabbit, blue/grey with a large ruff and the most protruding teeth I have ever seen in a rabbit. He is very old and plump, so must have been well cared for. He had, however, been left abandoned in a garden hutch and a kind lady took him into her home before bringing him here.

October 2nd 1990

With Rodney's help I have constructed new cages inside the Reception barn. Ollie is installed in the first one and looks very happy on his new perch with an electric blanket on the floor, the sort used for baby piglets. He is eating every wild mouse we catch for him. A tiny bantam died which he also ate. He later regurgitated a huge pellet showing that his digestive system is working normally again, just as it would in the wild.

October 4th 1990

Mrs. Schneider came to look at Ollie and adopted him for £5. We finished the rustic-clad cages in the Reception Barn — they look good. There are eight, plus a larger one for chipmunks.

October 7th 1990

Gale force winds ripped apart the peacock netting on the flight pen. I'm very worried about it. With Wendy's help we draped over an old ex-Army camouflage net. I rushed into Banbury to buy two more from *Troopers* in order to cover the other gaping holes, a very difficult task in the howling wind and torrential rain, — but we did it! The peacocks and pheasants are still with us; we were just in the nick of time. Only one escaped.

October 8th 1990

Sunny Sunday, lots of visitors, very morale-boosting. So different to yesterday, what a difference weather makes to our visitors.

October 9th 1990

I received a substantial cheque from Hook Norton Rural Fayre organisers, Jan and Roger Hughes, on behalf of the committee, – a wonderful boost for the Sanctuary. I shall spend part of it on some new signs at the entrance gates, the rest on seats.

I transferred the chipmunks into their new cage, but they found a tiny hole in one corner. Three escaped and there was much fun and games before I caught them again. I wonder if they will find any more pencil thin holes to squeeze through? They seem to love their new environment, all woodland-like with lots of natural bark to nibble. They are very happy scampering up and down, back and forth and are very friendly little animals and much loved, as they regularly take sunflower seeds and peanuts from visitors' fingertips.

OCTOBER 12TH 1990

Rod and I are creating a cabin inside the Reception barn, using eight feet by four feet frames side by side. Roger kindly brought four large sheets of toughened glass, perfect for the fronts. He was going to take them to the waste tip, but they were just right for the job here.

OCTOBER 20TH 1990

A telephone call from Clive the vet. Did I want a grey squirrel that had just been brought in to their surgery? I agreed to take him; he had been concussed, but when he arrived here he was snapping and snarling in the cage. I then let him out by our magnificent old oak tree and up he went to the top, full of the joys of life. I have observed one other grey squirrel hereabouts, so with a little bit of luck there might be one of each and we will then have our own small colony. They are delightful little creatures, so long as they are not caged. Although of a similar species to our tame chipmunks there is no comparison in their natures. Some chipmunks are hand-tame, unafraid and quite willing to be handled without attempting to bite. Visitors are fascinated by their antics as they are never still for more than a few moments. They eat all kinds of seeds and some hedgerow fruits gathered around the Sanctuary.

OCTOBER 21ST 1990

David Wilcox and his sweet wife came to deliver the piglets we had chosen two days previously. They are pedigree Oxford Sandy and Black, costing £25 each plus £3 for tattooing the female — quite delightful and friendly and happy in their new pigsty and run. I gave the male, called Winston, away to Katie who has been a volunteer helper. The female's name is Winniepiglet.

OCTOBER 23RD 1990

Ollie owl is fading now, and has stopped eating.

OCTOBER 24TH 1990

Ollie finally died after enduring several weeks of pain as a result of his injuries. I had hoped at one stage he would make a complete recovery. Perhaps it is for the best, as we would have had to go on feeding him in captivity and he would never again know the joy of freedom. On close inspection we found his claws to be quite amazing. There were little pads on each toe, resembling sea anemones. The clusters of small suction pads ensured live mice or voles caught in the wild would stand absolutely no chance of escape from those talons. He is now in the deep freeze along with the other collection.

OCTOBER 28TH 1990

A nuthatch was brought in by a Hook Norton man. It had flown into a glasshouse and was instantly killed, quite unmarked: it is a perfect little specimen.

OCTOBER 29TH 1990

Plentiful rain now which is filling the ponds; the grass looks a lovely shade of green again.

No Hunting Here!

A small boy with bright red hair visited with his grandparents from Birmingham, and held little mice who tickled his hands. He giggled a lovely happy giggle and kept on giggling, – music to my ears. His gran said she had never heard him laugh like that ever before. The baby rabbit likewise: he held this tiny mite close to his chest; it wriggled and tickled him and laughter was filling the barn. What joy, oh what joy! After they left we heard hounds barking very near the Sanctuary.

I rushed out the back and there was a whole pack of filthy hounds in my back field. I ran across the field and when close enough to the Huntsman, (Red Coat), I screamed at him to get them off my land, several times shouting *"call them off", "call them off."*. I dislike them intensely, especially the dogs. The barking and baying of hounds really sends a chill through my whole body, bringing back traumatic childhood memories of being severely bitten on the face at Fencott by a neighbouring farmer's dog, so I was very angry when the Hunt allowed their dogs to race through my Sanctuary. All the birds and animals were terrified. Sheep, goats and our little Shetland pony had been grazing so peacefully a few minutes earlier. I had previously insisted that the Hunt should not cross my land and had had many negotiations with the Hunt Master. I am adamant that they should not disturb my stock. Luckily there were no visitors just then and the Sanctuary gradually settled back to peace and tranquillity. The Hunt Master apologised.

OCTOBER 30TH 1990

Sheep dipping time, fun and games catching up the Soay sheep. First we captured Zeb the ram and secured him to the gate. Finally the others were caught, safely penned in, and we could get on with the job!

NOVEMBER 8TH 1990

We have noticed quite an amazing thing. Since the electric fence has been on all around the three acres of the Sanctuary, we have not smelt fox smells within or around. Foxes seem to know by experience that if they come too close they will receive a sharp sting on the nose from the wires; it does not kill them.

Still working on divisions to the pens inside the peacock pen, dividing into six breeding pens for the spring. After just writing that fox seem to have learnt to stay clear, I was damming up a pond and missed some Indian Trout runner ducks. There are only two left in their pen, six had disappeared. There were no feathers around, so it could have been two-legged culprits. I reported the theft to the Police. I did, however, fix an electrified wire above the roadside fence where there was evidence of human entry.

NOVEMBER 10TH 1990

A glorious day, sunshine, no wind: I was completely alone for once, so decided to take a break. It is a very rare thing for me to do and I must do it more often! A tiny rabbit played happily beneath the observation tower as I walked quietly and tip-toed up the steps to the top. Looking down, I watched as mother rabbit grazed, and the babies, one by one, played around her unconcerned. They lay on their little backs suckling her warm life-giving milk. She, (wild), had been rescued last year by a caring member of the public and is one of many fostered by a domestic rabbit. She is quite unafraid of humans, provided they are quiet and walk steadily. Squirrels dashed up and down round about and then down to the ground foraging for acorns, one of them being the squirrel brought in earlier. A moorhen stalked delicately across to the nature pond. I could hear the distant sound of goldfinches and many sparrows and yellowhammers. A quiet period to cherish.

Roy Invine has been back for two weeks, building a cow shelter. Rodney is hauling off-cuts from Eynsham Sawmills, six tree trunks from Bruern Estate office, £8 each. The cow shelter is finished in ten days at a cost of £600, far cheaper than a conventional barn but obviously not as durable.

Mexican Advertising Film

NOVEMBER 14TH 1990

An advertising film is to be shot here for a Mexican bank gold card promotion. I have agreed to line inside the peacock enclosure. Five rolls of close-mesh wire were brought by Ben Honeybourne for this.

NOVEMBER 15TH 1990

Rod and I continued with the netting lining inside the peacock run and making a cage for catching the birds after filming. The remainder of wire came, then on with the job — quite difficult as we must not leave any small gaps through which the birds can escape. Doing battle with wind and rain is no joke! It was almost gale force at one stage today

Five days from the theft of Trout Runners, the police came to investigate. Archie from Hook Norton is writing about the theft in the newspaper, but I do not hold out much hope of recovering them.

NOVEMBER 18TH 1990

I went to Norfolk today to buy a pair of Coscoroba swans from Brian Boning. They still have some grey feathers as they are not much more than cygnets, born in June. They have bright red beaks, red feet and are extremely elegant. One named Daniel is named after the little boy who, level with the counter, questioned if I had any Coscoroba or Trumpeter Swans. I promised him that I would get some one day and today's the day! The female I have named Diane after Daniel's devoted foster mother. My Coscorobas were like Brian's babies and I don't think he really wanted to part with them. When you raise them by hand they become very tame. Wendy and Rod looked after my precious Sanctuary.

NOVEMBER 22ND 1990

The peacock run is ready for the film crew and on Sunday they are shooting a portion at Broughton Castle. Ben brought twelve cages from Crofts into which we will put our birds for filming their release into the covered area.

NOVEMBER 23RD 1990

Oxfordshire Special Conservation Award judges toured the Sanctuary and one or two seemed impressed. Reg Little said he would like to do a feature for his farming pages in the *Oxford Times* newspaper, about making good use of otherwise useless agricultural land.

NOVEMBER 24TH 1990

After dark, Wendy helped me pen cockatiels into the larger cages, budgies into the medium-sized ones and finches, zebras and Bengalese into the smallest cages. After filling up the feed and water containers we left them in the large aviaries to settle down for the night.

Thank you my duck!

NOVEMBER 25TH 1990

Ben Honeybourne arrived at 8 am today (Sunday) to take the small birds to Broughton Castle for the filming. Next Sue Ryan arrived with her wonderful spread of food to feed the twenty film makers. Two very large vans containing the generators and filming gear manoeuvred into the Sanctuary at noon. The scene was set up inside my peacock runs. Ten of the cages with birds inside them were carried very carefully into the large safe enclosure. Filming was expected to take five days on and off. Shooting began and the sun was shining – a really lovely day for the end of November. Massive lights were beamed on to the set, ensuring the film could have been made under any conditions, except rain or snow. The idea for the commercial was that a multi-millionaire's daughter wanted to buy all the birds in a pet shop, take them home (Broughton Castle) and then release them all into the sky to fly away free. It was pure artistic licence, of course. The reality was they would barely have survived a few hours, the wild birds making very short work of them. The first three canaries were released to fly free (inside the large enclosure) and I was to catch them up with a net as they were released. Off I went pursuing them to the far corner, and wow, down I went into the slippery slope of the duck pond, bump on to my bottom, slushy mud covering my feet and up to my bum. I had to change into clean clothes, so disappointingly I missed most of the filming. After completing outside, the crew came inside the Barn to eat. In the next scene the cages were placed around with more of my little birds inside to re-enact a pet shop. It was filmed from outside through my windows and across the large aviaries outside. All of this had been tried out in Woodstock, Rome and France without success, but our Sanctuary proved ideal. Everyone seemed very pleased and one lady member of the crew told me that when she started work this morning she had been in a foul mood, but by the time she had observed my ducks and pheasants, peacocks and hens so peaceful and happy, she too became at peace with herself. The film makers were *Geoff Reeves Films Limited*, London. The Mexican banker sat in my grubby back kitchen very happily eating *chilli con carne*.

Oxford Times article by Reg Little, one of the judges for the O.S.C.A awards *(see p.64).*

NOVEMBER 27TH 1990

Ben brought my cheque for £175, including £25 for Sue, fee for her food. He also left three of the cages for me to sell off later.

NOVEMBER 28TH 1990

A drake named Douglas was brought in, hatched from one of my eggs sold to the visitor in the spring. She also brought another large beautiful white duck, who was very tame and waddled around in the shop for a while before joining the others on the large pond. We carried on with the shelters in pheasant breeding cages. We fixed an extra length of wire, three feet by six, vertically inside two corners. It was then packed with straw and side-stitched up with the help of a work experience lad, making a cheap, but effective temporary windshield.

DECEMBER 6TH 1990

Saturday. Deep snow — and I couldn't get out of Enstone. Rodney managed to get there early by tractor. He assured me everything was OK by phone, luckily still working. He managed to catch four Brecon Buff geese and house them safely despite the raging blizzard and was wet through to his waist. Electricity was all off at Wigginton and Hook Norton, therefore no fences were working against the foxes. They have been around but kept a distance from the electric wire. How clever they are, having learned not to get too close.

DECEMBER 7TH 1990

After a sleepless night worrying about everything, I managed to get over to the Sanctuary with difficulty. Den, Wendy and I assessed the damage: the water was all off but the telephone was still working. The peacock pens have partly collapsed through the weight of the snow on top of them, so we sheared them up temporarily.

DECEMBER 8TH 1990

Snow thawing, thank God.

DECEMBER 9TH 1990

All services back to normal again.

Conservation award

Wildlife worker wins award

ANIMAL lover Mabel Warner has won an award for her conservation work.

Mabel single-handedly runs Wigginton Wildfowl sanctuary near Banbury which is home to hundreds of birds, rabbits, goats, pigs and ponies.

The award was one of 10 given for the first time this year by Oxfordshire County Council for people who have made an outstanding contribution to protecting the environment.

"I was extremely pleased to get the award and am going to mention it on my brochures from now on," said Mabel.

Ian Walker at Oxfordshire County Council said the awards were introduced to promote and increase awareness of conservation issues.

"The wildfowl sanctuary won an award because of its contribution and commitment to conservation," he said.

DECEMBER 18TH 1990

I won an O.S.C.A. award for outstanding contribution to conservation in Oxfordshire, thanks to Wendy sending in an application form. These are the first awards of this kind ever to be held in Oxford. The chairman described our Sanctuary as unique, an honour indeed, and one of the ten winners out of forty-six entries. I feel our efforts are being appreciated by many people and I trust this award will encourage many more visitors in the future to come and support the work we are doing at Wigginton. We shall now be able to feature award details on our leaflets and brochures.

CHRISTMAS DAY AND BOXING DAY 1990

Rodney managed all of the feeding, letting out and shutting up at night, with a little help from Wendy. I took the day off and went to see my mother who is nearly ninety. Dennis cooked Christmas dinner, bless him, then we all enjoyed our meal. Praise the Lord, a lovely Christmas Day! On Boxing Day a gale force wind blew a tree down across the Tew road, just missing Wendy driving my van with her two children on board. God was certainly with her then, as he is every day. Two brave visitors only.

Young artist Joanne Arber from Great Holm, Milton Keynes sketched these in 1990: Angora rabbits & Canada Goose

JANUARY 1991

January is a horrid month with ponds freezing, pipes freezing and water-bowls freezing.

Two seagulls brought in, blown by the gales. Guinea pigs and baby rabbits brought in also. A lady rang from Milton Keynes who had lost her Campbell duck and had looked everywhere for a replacement and then was told about us. Yes, we had a youngster from our 1990 hatching!

Nick Allen from Adderbury brought in the proof "Map of the Sanctuary", looking good, £144 for 2000, A3 size. He is a kind man.

An invitation to a craft fayre near London, three days in early May. I might go and display some birds and animals.

JANUARY 8TH 1991

Four more young rabbits brought in and four Khaki Campbell drakes. I saw an article in the *Oxford Times* dated December 28th by Reg Little, as he saw it on the day the film crew had left. Like most newspaper writers he couldn't get all the facts exactly right. The description of the little girl in the film who bought the caged birds and then let them out to freedom is slightly different to Reg's description. It was done very tenderly and with much feeling for the birds. She had seen them locked up in small cages inside pet shops and wanted to let them all free.

January 12th 1991

A tearful Saturday morning. A fox ate one of my Coscoroba swans, whether male or female we cannot yet tell. Wendy thought Rodney had shut them in and Rodney thought Wendy had shut them in, but alas, neither had. I had really begun to think fox was leaving us alone.

January 13th 1991

A nice few visitors, a sunny day: three more rabbits brought in and another five ducks — I must try to re-home some of them soon.

Coscoroba Swan

January 18th 1991

Wendy is a welcome help at the Sanctuary, working alongside Rodney. I moved peacocks to roomier pens, pairing them off (hopefully). One male escaped, flew off to the old railway embankment, and perched up in a tree all night. Next morning he came back but flew once more to the top of the oak tree as we tried to capture him. This time we decided to leave him loose. I would love nothing better than to see all my birds flying free, so long as they stay perched up high at night to escape foxes. One particular large black turkey flies out of his pen each day, but returns in the evening time to be shut up safely.

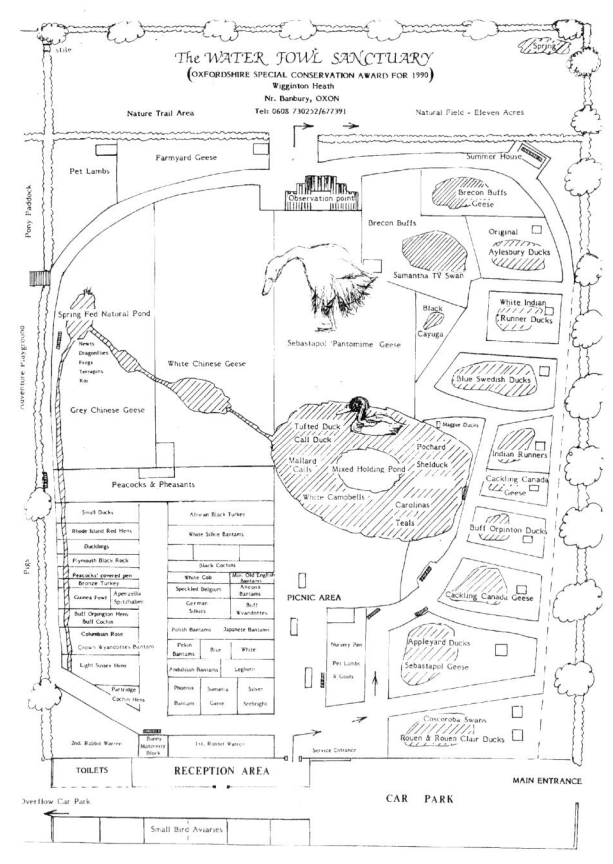

Nick Allen's map

January 30th 1991

Nick Allen, the artist from Adderbury, brought the Sanctuary maps he had kindly made. They are beautiful with all information typed on the back. They should sell well; he's a good friend to us.

February 8th 1991

Severe icy weather, pipes frozen, ponds frozen, dishes, water drinkers all frozen each day. We had to give birds small plastic bowls of water, keeping us busy constantly. Down to minus 13°C, the coldest winter for four years and the forecast is for more snow to come.

A little owl and a barn owl came from the vet's surgery in West Bar. The barn owl has a dislocated wing, the little owl was concussed.

February 15th 1991

A little grebe brought in, rescued after being found frozen stiff in a garden at Tackley. He is a wader bird some distance from water so I presume he lost his sense of direction. Snow-covered ground seems to have this effect. I have noticed my hens running free range; they are reluctant to venture very far from their houses, and if they do, they seem to be disorientated by the carpet of glistening white snow. An old country saying is *'silly as a chicken'*; since raising many hundreds of rare breeds by hand I find quite the opposite to be true.

FEBRUARY 19TH 1991

Sparrowhawk found, unable to fly brought in, a beautiful bird, with no apparent sign of injury, so I hope he will make a speedy recovery (not eating at all.) Little owl and barn owl both doing well, eating and pelleting normally, the barn owl now in large aviary, ready for release back to the wild. Farrier Mark Johnson came to trim Penny, the pony's feet; this needs doing every few weeks. She suffers laminitis if she eats too much grass. This is painful and must be avoided if possible.

FEBRUARY 25TH 1991

A few brave visitors at weekend; they seemed to enjoy it in despite the wet and the mud. Thomas the goat starting getting out again. He and Danny are nuisances, I would like them eventually to go to good homes.

Sting on the Nose

FEBRUARY 28TH 1991

Marcus our next door neighbour brought back my electric fence unit which I had lent to him earlier. We then proceeded to fence around the little top paddock to include a house, putting round three strands of wire, we connected it up, it worked well, so gathered up the goats Danny and Thomas, Wendy holding one, me the other, and then released them. Straight to the fence they trotted, then a sharp baa-aa from Danny: his sting on the nose, soon taught him to stay clear. But Thomas, – oh no not Thomas – he ventured again and again towards and through the wire, despite being stung. He tangled himself in the wire, butting and jumping, and pulling out the stakes as he went. That was no use, so we put yet another strand around, making four in all You've guessed it, he went straight through again, regardless of the electricity. Will he ever learn? Next we put another row of sheep netting inside the strands and connected up once more, then left them both for the night hoping they would lie down in the house and behave themselves. By now it was quite dark. Bingo, Danny and Thomas are at long last safely behind an electric fence and there'll be no more time wasted on them I hope, at least for now!

MARCH 2ND 1991

Dorothy's friend, Michael, from Banbury, brought his two little ponies to the Sanctuary, Whiskey and Vienda: mother thirteen, daughter eleven. Whiskey the mother is smaller than the daughter, who is darker in colour and is also fond of nipping sleeves and coat edges – until she gets a titbit, that is. She has been thoroughly spoiled by Michael, who loved them both dearly and is feeling unwell and ageing. Rather than sell his beloved ponies to a horse dealer, knowing full well what would happen to them, he brought them to me, safe in the knowledge they would be well cared for here at the Sanctuary. They make another added attraction for our increasing number of visitors. Just one more paddock to be created, preferably including a water-hole as Vienda loves nothing more than a roll in a pond, would you believe?

A small jackdaw has been brought to us with a broken wing. He was soon fixed up and is eating well, – a perky little bird is Jack.

MARCH 5TH 1991

Wendy noticed that my large sign had disappeared from the side of the A361, Chipping Norton to Banbury road. On inspection I found that the other large one had also been removed and taken away – by travellers, I have strong suspicion. The Police and the sign-makers say the same; as they are made of aluminium, they steal anything and sell it for scrap, it's disgusting to think they can get away with such crimes. All replacements will be made of plastic.

Councillor Mr. Mike Cowlan came to say I **must** get the gateway done. I'll have to get estimates, — another big shock no doubt.

MARCH 14TH 1991

Little owl brought back by Rodney found in Marcus Hughes' barn with damaged wing, but still bright and cheerful; he had escaped out of a tiny gap in the wire. Constructing large peacock pen alongside the other one for our seven youngsters that we hatched last year as they are now beginning to grow longer tails and need more room. Sold two ducks, three geese including my little one disliked by most of the others. They all pecked her and made her life miserable. Then when she went to Jan of Hook Norton she quarrelled with her geese; maybe she just doesn't like being penned in small areas with others. This man who bought them, however, keeps red deer in a large fenced area of woodland with a stream running through it, so my baby goose should be very happy in that kind of environment. I do hope so.

Baby chicks are hatching now: Black Cochins and Bantams. Many baby rabbits for little visitors to cuddle at Easter. Honey, my Cashmere buck, had fathered some lovely offspring this springtime, mated up with the small dutch honey and white. The tiny doe, the colour of a Siamese cat, has six sweet babies including one very tiny perfect little specimen.

MARCH 15TH 1991

I moved two Cockatiels over to large aviary as they were being pecked by others. Moved Turkeys to a clean patch once again.

MARCH 16TH 1991

Attended a farm sale at Dunthrop Farms; Nancybell Gregory, a radio personality, previously farmed there. I bought Weldmesh wire, a Shepherds hut on wheels for the Children's Playground; pallets, buckets and lots of bargains.

MARCH 17TH 1991

Small baby ducks hatching, more chicks, only three so far. More Terrapins brought in. I once again spotted frog spawn in the goldfish pond, – very pleased about that.

MARCH 18TH 1991

We finished off the other Peacock house and run to move them out of the aviaries and give them more room. Wendy is painting aviary doors; good girl.

MARCH 27TH 1991

A nice few visitors for Thursday, lovely weather.

MARCH 29TH 1991

Good Friday. Sanctuary's 2ND BIRTHDAY.

MARCH 30TH 1991

Sold my first Cashmere baby rabbits £10 each.

Sanctuary cash burden a threat to animal lover

By JANET McMEEKIN

THE award-winning owner of an Oxfordshire sanctuary may be forced to sell off hundreds of animals to clear her debts.

Mabel Warner, 60, has been funding the sanctuary at Wigginton Heath, near Banbury, out of her own pocket since it opened more than two years ago.

But over the winter she has been struggling to feed the 2,500 animals including waterfowl, ponies, goats and sheep, because the public have almost stopped visiting.

"It's fine during the summer, but when the colder weather comes everyone tends to forget about the sanctuary," she said.

"It's not a charity and because I'm having to pay for all the food, I'm running into thousands of pounds of debt. It's getting to the desperate stage".

Mrs Warner says that although the 22-acre sanctuary is in no danger of closing, she may be forced to sell some animals.

"It would be great if people would remember about us at this time of year and come along to help boost our funds again," she says.

Last year Mrs Warner received a conservation award from Oxfordshire

March 31st 1991

A Tourist Board representative came with her son was thrilled with all of it. Many visitors, all looking very happy — I had to use the overflow car park, a lovely day.

April 1991

Leaflets from Cherwell District Council including four evening visits to the Waterfowl Sanctuary throughout the summer at reduced price of £1 per head instead of the normal price of £2 per adult.

April 6th 1991

Birthday party. Our good friend Sue Ryan and daughter Victoria laid on a wonderful spread, but the wind blew and rain came pouring down all day so there were not many visitors.

Nick and Joy Allen came, I gave them a bottle of wine as a thank you to Nick for making the maps that are proving so popular: they sell well at 50p or £1 laminated ones. Thanks to Sue and Bernard Taylor, of Middle Barton, for laminating at a fraction of the right price, and encapsulating my photos with their off-cuts out of the goodness of their hearts. Little Abigail is a frequent visitor. My best friends have free access to my Sanctuary indefinitely. Joyce Holt and her husband, John, came and brought me a rubber stamp that they had made specially for me. They run a craft business, making and selling rubber stamps. My other good friend, Margery, now uses these stamps on her cards. Michael Page, YFC Secretary, came for a few more eggs: 2 Wellsummer, 2 Buff Orpingtons. He's getting a fair hatching so far, mainly concentrating on Marans, he had one of my good cockerels. He asked me to put on a display of rare breeds at Middleton Stoney YFC on May 18th – I look forward to that.

April 7th 1991

A man came down from Peterborough for some fertile goose eggs.

April 8th 1991

I'm sixty today, an OAP, and I feel 40.

Lambkin had given birth to twin black lambs in the night, Wendy cried tears of joy at the sight of them first thing this morning. Farrier finished off ponies feet; they are all comfortable now for the next five weeks. It's lovely up in my top fields. Two Peacocks out on the road, three passing motorists kindly told us of their escape outside. I ran up to the top of the hill after them, and they flew gracefully back down to the bottom! Silly me!

April 14th 1991

A little American boy called Evan came in and said *"Oh I do like your chicklets"* — that sounded so sweet to me, I asked him to write it down in the visitors' book.

Two baby ducklings hatched and many more chicks, all rare breeds.

I have re-homed the last of the farmyard ganders; now I would prefer to be concentrating only on rare breeds of geese but I am reluctant to turn anything away.

Very stormy weather, laying out posts for the fences out in the back field.

Stanley Cox from Swerford came to say a fox was taking poultry from his back garden and asked if I would go to the rescue of the remaining five hens – two ducks and two geese. Monday evening as arranged I went from 8.30–9.00 p.m., collected the hens and ducks to bring to the Sanctuary's safe haven, next evening I called at 8.15 p.m. for the geese. I could find no one around, so I went up his garden, collected them and took them to Wigginton. I returned at 9.15, still couldn't find Stanley, but a light had come on upstairs, so I assumed he was OK, but he had left his back door open,– a bit worrying, as he was very upset about his poultry losses to the fox.

April 16th 1991

Stanley's geese are now sitting well.

A lady arrived with a goose house in a trailer, 3 geese and 2 hens; she gave £50 donation to buy more wire fencing. Very much appreciated.

April 20th 1991

Twin Soays born — we got them safely into a shed after much ado.

April 22nd 1991

Twin Soay lambs doing well. Lou Lou seems to be calming down a bit. Soays are so very highly strung, nervous-natured little sheep. We lost one young Peacock followed shortly by a Peahen: they seemed to become lethargic for a couple of days, went off their feed, and then died very quickly. The ones flying loose are fending very well for themselves and looking quite beautiful. Three more geese brought in, two of them rather troublesome: Doris, Oddsey and Beaker from Wardington.

Crows and magpies are taking a toll of eggs if laid outside; more chicks hatching, blue Pekin, black Orpington, Wellsummer, Phoenix.

Main sponsor Alex Potter (right) with friends

Telephone call from a lady at Bloxham to say her little girl Alex and two friends had done a sponsored litter pick in aid of Sanctuary funds. A photographer arrived to take pictures of the three girls cuddling our two baby Soay lambs born last week. They are so sweet. The photo is to appear in the *Banbury Cake* newspaper next week — more good advertising, thanks to three wonderful little girls.

April 27th 1991

Young golden pheasant brought in from Irene at Little Bourton, – very beautiful; also a pair of white peacocks, even more attractive, their age unknown, were brought from Southam Zoo, (which has now closed down). Peacocks originate in India.

May 2nd 1991

Wendy away on lambing course. Typically, Larryadne (Wendy's pet lamb) decided to give birth to one black lamb this afternoon! I was here alone, heard her heaving; went to investigate and found its head coming first, no sign of the feet. I hadn't the strength to help, and thought the lamb was dead. So I called the vet, who said she would come at 5.30; she didn't arrive, so I called up Judy from Manor Farm next door. As she arrived, so did Wendy and they managed to help the birth; Judy got it breathing, then handed it over to Larryadne, who proceeded to clean her new born very lovingly. The vet arrived, inspected it, said what a super big lamb it was and sprayed its navel for quick healing. I was happy that all ended well after my panic and not knowing what to do, – relief all round.

May 3rd 1991

Wendy took one of the other twin lambs into the vets as it was scouring again. An injection cleared it up quite quickly. Our nice strong Soay lambs resemble baby fawns and I'm naming them Lindsie and Laddie. The little boy had to have a castration ring fitted as we don't want another Zeb yet.

MAY 4TH/5TH 1991

We went to the show at Moredon Hall Park, London, the result of winning a Conservation Award for Oxfordshire 1990. John and Joyce Holt, my dear friends from Adderbury, took me in their car.

We took one cage divided into six compartments with Blue Pekin Bantams and chicks, Black Cochin and chicks, a mother Angora rabbit and babies. I sold one baby Angora the 1st day, the 2nd day was as cold and wet as the first, so I decided I could not manage a 3rd with the excruciating pain in my back. The weather was wet: an east wind blowing into the open tent, taking the pleasure away. The children really loved cuddling the babies, amazed and delighted actually to be allowed to handle them; their faces were a study. I wish I'd had a video camera – maybe one day I'll own one.

MAY 6TH 1991

Today a lady called May rang to say she was coming from London to buy the baby blue Angora rabbit she had seen at the show. She found us quite easily with the map on the back of the leaflets. I've handed out quite a few; time will tell if any good comes of it.

MAY 9TH 1991

A young man brought in Buff Cochins, Partridge Cochins and Dark Brahamas;, beautiful birds with feathered legs, they look like they are wearing trousers.

An Indian White Peacock makes his beautiful display

MAY 10TH 1991

A gentleman came and bought four of my young chipmunks for £20 each, and exchanged one of my last year's males for one of his young ones to introduce new genes.

PRIME TIME
Lifestyle focus for the 50 plus

A Goodhead Publication. No.5 April/May 1991 Telephone: Oxford (0865) 880505

One woman's determination sees a dream come true

MABEL'S ANIMAL MAGIC...

Retired florist's wildlife haven

Report by PHILIPPA SODEN

MARVELLOUS Mabel Warner created her own animal magic when she gave up everything to realise her lifelong ambition of owning a bird sanctuary.

Two years ago she handed over the running of her successful flower shop in George Street, Banbury, to her daughter and sunk all her life savings into 22 acres of marsh land at Wigginton.

She and husband Dennis moved out of a large comfortable house into a one bedroom bungalow and took out a huge loan from the bank.

But Mabel, 59, from Enstone, says she has no regrets and is happily looking forward to the future with her Wigginton Wildfowl Sanctuary.

"I owned flower shops for 25 years but my eye sight started to fail and I could not see whether the flowers I was buying at Birmingham market had spots on or not," she said.

"It has always been my dream to open a bird sanctuary. It cost me everything I had but I have no regrets."

The sanctuary is open seven days a week to visitors and Mabel is always coming up with new ideas to encourage interest in wildlife.

Her latest idea is 'teacher packs' for local schools to encourage teachers to do in-class projects combined with visits to the sanctuary.

Her efforts were rewarded this year when she won a special award for her conservation efforts from Oxfordshire County Council.

Mabel runs the sanctuary practically single handed, although she does have help from husband Dennis, who is semi-retired.

He does all the heavy work and has Mabel's dinner on the table when she gets home at night.

"I could not do it all without him, he is a real gem," said Mabel.

Married

The pair married five years ago.

"He was my childhood sweetheart, we used to go out together when I was 17 and I always had a soft spot in my heart for Dennis."

The sanctuary started with just a few wild birds but has now grown to be home to

★

'REALLY WILD: Mabel Warner gave up her nice home to virtually single handedly create her own 22-acre wildlife sanctuary that has been presented with a special award for conservation by the county council

★

over 2,000 animals.

People now take animals they find to Mabel so she can cure them, before releasing them back into the wild.

May 11th 1991

Joyce Holt gave a substantial donation to help buy food for the Sanctuary, (or more wire), as a proportion of their craft takings during their recent successful three days at Moredon Hall Park.

May 12th 1991

One Sebastopol gosling hatched. Baby Canada gosling doing well, brought in recently; baby white Chinese and two Brecon Buffs are fine. East Indian duck sitting tight for seven days; I transferred some of her eggs to an incubator.

May 16th 1991

Saturday, an early start to the Young Farmers Club Show at Middleton Stoney — a very successful day; I gave out lots of leaflets, met lots of interesting people, £50 sponsorship for the Sanctuary. Rodney manned the reception at Wigginton today and had quite a few visitors in spite of the wet weather.

May 19th 1991

Went to Stowe Fair. I bought a second-hand incubator and six strong water drinkers for my hens and bantams.

First golden pheasant egg hatched.

May 22nd 1991

Nine goslings so far, many chicken, rare breeds: everything looking very happy at Wigginton thanks to Wendy and Rodney.

May 25th 1991

John Ekers from Bloxham College borrowed a Cochin hen and some young chicks, also two goslings; he displayed my maps, distributed leaflets, and raised £20 for the Sanctuary. This helped their open day with Conservation Studies – good from an educational aspect. We have had baby rabbit success with Mabel's Midgets: eight young does are nearly ready for breeding — helping to produce a friendly, loving, cuddly generation of little rabbits not growing too big for children.

May 26th 1991

Bank Holiday Monday — a lovely day: many visitors: baby rabbits are definitely favourite with the children. One little boy about ten years old was visiting with his Mum and Aunty in the morning, and fell in love with one dusky brown Angora Seal Point Siamese colour rabbit. He planned to go home to London to ask his daddy if he could have it. Two more from the same litter were booked.

Peacocks are displaying beautifully to all the visitors. All young males fly loose now around the Sanctuary just as I'd hoped they would be able to. At night they roost high up in the Oak and Ash trees, safe away from foxes.

One of Michael's horses bit a small boy's arm – no blood was drawn, but it frightened him a little; I wrote to Michael that they must be removed immediately; I thought Vienda had got over biting as I'd spent nearly £100 on Vet's and Farrier's fees, but no, – so they must go. I can only keep gentle animals and birds for the public to be near.

May 27th 1991

The little boy from London came back in the afternoon with his daddy's permission, paid a deposit to secure the baby bunny when it is weaned from its mother on June 16th. Joanna, the little London girl, also came back, bringing her parents this time. I helped to make this little girl happy last year by allowing her to hold baby chicks and then by enabling her to own one in London; apparently the chick wasn't quite as happy in London as it was in the countryside.

About Olitwo. A lady from Farnborough Hall rang to say she had found the young tawny owl, no sign of its parents. The bird must have fallen from a nest somewhere. She rang the RSPCA who told her to bring it to my Sanctuary, but as she was coming to West Bar she kindly left him at my very helpful Vets, Clive Madeiros Veterinary Hospital. I picked him up next day, – sheer delight with those huge round eyes gazing out of a round fluffy face, still with the baby down, a few tail feathers starting to appear, and also wing flight feathers. He's eating every wild mouse that we can catch for him, we have many traps set for the wild mice who become quite a nuisance in the aviaries, nesting in the nest boxes and consuming expensive seed. Olitwo is growing well, he clicks his beak when approached turning his body upside down, his claws are menacing and he has great wide-open eyes. He's very amusing this little baby is, and oh so beautiful! He has a heat lamp at night to keep him warm. Last night he'd scrambled out of his cage and somehow got into a feed sack standing nearby with a little bit of corn still in the bottom. I think he thought that was the next best thing to a hollow tree, and felt safe inside.

Dippy

JUNE 2ND 1991

Following an invitation to a Rural Fayre in Shipston-on-Stour, I went along with my Pets' Corner, taking just my small Bedford van containing rare breeds of chickens, ducks, goslings, also baby rabbits, and everyone's favourite: my huge white Angora rabbit called Snowy. She is so docile; she just sits or lies full length on a table top, allowing everyone to stroke her beautiful fur. She has a thick, fluffy coat and she grooms herself continuously keeping it in extremely good condition. Snowy is much photographed with her large beautiful pink eyes and the tufted tops to her ears. She's not interested in having babies like her sisters; she's the film star of our family. Also I took a small very tame chipmunk called Dippy; he promptly escaped from his cage through a small feeder hole that I'd forgotten to cover up, this being just a show day cage and it was the first time I had taken a chipmunk to a show. Anyone who is familiar with chipmunks will know how fast they can move, being a type of squirrel from Siberia. It was not too easy to catch him; imagine lots of well-wishers, men and women darting about in and out of the parked cars, in one of which sat a man holding a small Jack Russell dog at the open window. A yell went up *"Hold that dog!"*, as he would have caught little Dippy quicker than anyone, but it most certainly would then have been a dead Dippy. However, my friend Margery caught Dippy first, and promptly returned him to his cage; then he immediately got out again, and everyone was back to the chase, this tiny creature no more than six inches long from its head to the tip of its tail causing so much fun and excitement among grown-up people. It was lovely to see the expressions on their faces; not many of them knew what a Chipmunk looked like, with their pretty striped bodies of various shades from brown to cream, (there is also a white version). Like Mercury they are very quick movers: they almost fly through the air from one point to another and on the ground they resemble a

Dippy

Margery feeding her friends

Thank you my duck!

jumping jack firework. This episode was followed by many interested enquiries as to availability and the care of chipmunks. Two were later sold to a man with a pub at Newbold-on-Stour. They could certainly keep his customers amused!

Jacko

A tamed Jackdaw came back to us recently via the Police at Bicester. It promptly flew off to Milcombe, was reported to us, before arriving in Banbury, seeking out a chap who obviously loved birds, a breeder of parakeets etc. Jacko upset his birds so much that he had to go – so he was brought back to the Sanctuary. The man told us he will only fly on to the shoulder of chaps and not girls, but Archie, our local Policeman, said a Jackdaw had flown on to his daughter's shoulder as she was cycling along the Hook Norton road towards Milcombe. I expect it was the same bird. It is very friendly and loves a Policeman's shiny buttons and badges. We gave him some old shiny keys to play with; we kept him in a large aviary for some time for his own safety.

We have incubated around two hundred rare breeds of chicks so far and there are more in the incubators. The second-hand one seems to be working well; it will do as a brooder for next year.

My grandson, Micah, gets a nasty surprise from naughty Jacko

Banbury Cake

The *Banbury Cake* newspaper sent Georgina, a photographer, to take shots of Jacko the Jackdaw. When Jacko was being photographed he flew on to my little grandson's shoulder and pecked his cheek just as the photographer snapped her camera. Poor Micah! He did cry a bit! He wasn't expecting that sort of treatment. Jackdaws have very sharp, black beaks; one can see how easy it is for them to break eggs open and eat the contents. Unfortunately they do cost us dear in eggs. The wild ones steal: they carry away any that are laid in view of their beady eyes. The Crows and Magpies are equally destructive, even taking baby birds, if they are not removed to a safe haven first. It is a constant battle with nature here.

At around this time three tiny little greenfinches had been brought in, rescued from cats who had already killed one of their number. Wendy went into Paws Pet shop in Banbury to buy a packet of *Sluis*, a wonderful new food manufactured in Holland from natural airborne creatures. It helps save baby fledglings' lives, a wonderful development.

Caring hands

Banbury Cake

June 16th 1991

A baby sparrow brought in, saved from a cat; one baby Greenfinch died.

The little boy from London came with his Mum to collect his rabbit. His Mum commented that it proved a very expensive rabbit with all the travelling costs, but they did want it as it was so special and so tame.

Two fledgling little owls brought in, one died, the other is quite strong. Owls look at you in the most appealing fashion. Olitwo is gaining strength daily now. Anthony has started to hand train him for flying; he is very beautiful, growing his main tail-feathers and losing his baby fluff.

Three very rare miniature Poland bantams hatched, (they are so called because of the pole on the top of the head); they all have lovely white top-knots, resembling ready-made hats. The photographer Georgina sent me some photos, including the one of the greenfinches *(on page 75)*.

June 22nd 1991

I took a Pets Corner to Spiceball, Banbury, to help Chalkwell raise funds, and also to publicise us.

The nuisance pigeons, who insist on roosting in the Reception Barn area, and making such a mess, we removed to Enstone, and put them into the Tythe barn; only five stayed, the remainder were back at the Sanctuary before I arrived next morning! We wired over their favourite perches to discourage them They have plenty of other shelter places in which to perch and nest.

June 23rd 1991

Another Pets Corner at Freeland School Fun Day. The animals: baby rabbits, chicks, ducklings, all so popular with the children, who love to cuddle them: more good publicity.

June 24th 1991

A drake came in for a wing trim. Two Netherland Dwarf rabbits in for bad behaviour. Raining all day again: a wet, wet, June. I sold two pairs of Fantail pigeons, £10 per pair, (my babies); also three young Maran pullets – thank goodness we have some young stock to sell as there have been very few visitors because of the wet weather. I still have the feed bills to pay. One more Miniature Appleyard duck hatched; a little visitor saw it hatch out of the egg – a very special treat. She was brought to the Sanctuary by her Mummy as the rest of her class went on a trip to Birmingham, but this particular little girl got travel sickness on coaches and couldn't go.

June 26th 1991

Twenty-seven children came from Queensway School, Banbury. Zebbie, my Soay ram, butted one little girl very unexpectedly. He has never done that to anyone before. We shall have to separate him out of the nature field. I had already removed Thomas, the usual offender. Sophie, the little girl was unhurt. I took them up to the tadpole pond and let each of them hold a tadpole, cupped in their hands with a small amount of water. The tadpoles are just starting to grow their back legs.

Then ten boys with learning difficulties came with two teachers from Woodeaten Manor School, followed by the Over 60's club from Moreton-in-Marsh. All enjoyed their visits.

June 27th 1991

An evening visit was organised by Cherwell District Council. Fourteen turned up in spite of the rain, including an ambulance that, among others, brought Andrew, a young former Policeman who had been involved in an road accident, rendering him wheelchair-bound severely disabled. He thoroughly enjoyed his visit, as did all the others.

June 29th 1991

I took my Pets Corner to Deddington Primary School Fete. Once again it was very popular with both kiddies, parents and teachers.

Thank you my duck!

JULY 1ST 1991

A school came from St. Ebbe's, Oxford, and enjoyed it very much. Really they needed an extra half hour for their questionnaire.

Eight Highland Cattle arrived whilst I was at home nursing sciatica: excruciating pain in my leg and hip joint. The animals are very beautiful and enhance the Sanctuary with their presence. I am just grazing them for Mr. Tony Crawford from Tackley for the summer. They are all named after members of his family. I hope to be able to keep one or two for the Sanctuary in the Autumn.

Highland Cattle, Tracy and Rachel

Incubation

JULY 4TH 1991

A class of children came from Dr. Radcliffe's School, Steeple Aston, and walked all round the Nature Trail accompanied by teachers and myself – (with a walking-stick!) – quite a long walk for little ones, but they all seemed to enjoy their visit very much. We went close up to the Highland cattle; they are indeed magnificent creatures, very docile. The school borrowed an incubator and duck eggs; below and on the following pages is the record of their class project.

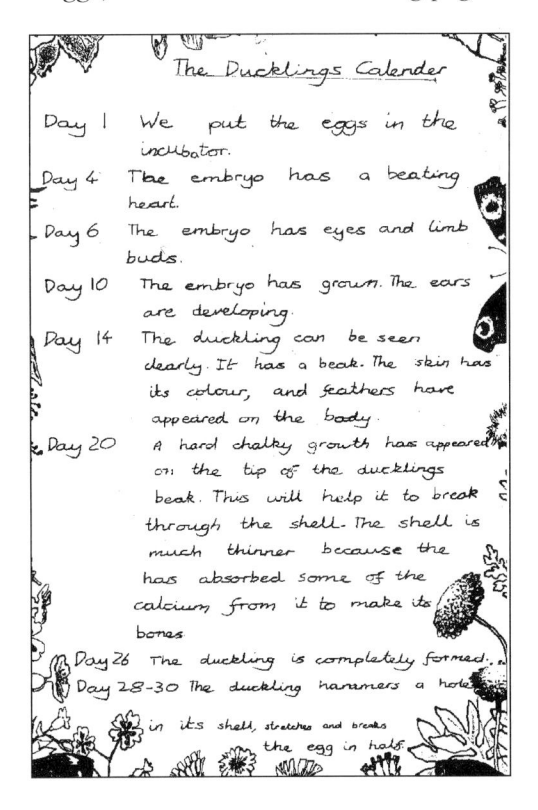

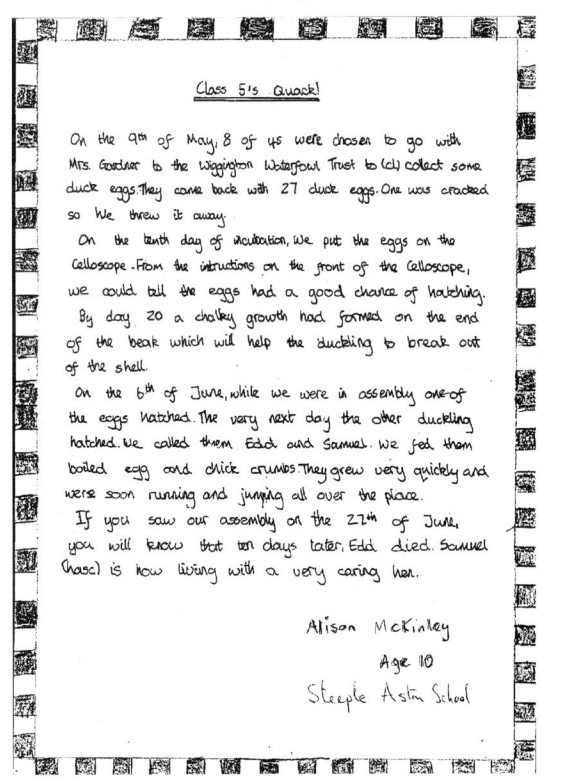

The Ducklings Calendar

Day 1 — We put the eggs in the incubator.

Day 4 — The embryo has a beating heart.

Day 6 — The embryo has eyes and limb buds.

Day 10 — The embryo has grown. The ears are developing.

Day 14 — The duckling can be seen clearly. It has a beak. The skin has its colour, and feathers have appeared on the body.

Day 20 — A hard chalky growth has appeared on the tip of the ducklings beak. This will help it to break through the shell. The shell is much thinner because the has absorbed some of the calcium from it to make its bones.

Day 26 — The duckling is completely formed.

Day 28-30 — The duckling hammers a hole in its shell, stretches and breaks the egg in half.

Class 5's Quack!

On the 9th of May, 8 of us were chosen to go with Mrs Gardner to the Wiggington Waterfowl Trust to (c) collect some duck eggs. They came back with 27 duck eggs. One was cracked so we threw it away.

On the tenth day of incubation, we put the eggs on the Celloscope. From the instructions on the front of the Celloscope, we could tell the eggs had a good chance of hatching.

By day 20 a chalky growth had formed on the end of the beak which will help the duckling to break out of the shell.

On the 6th of June, while we were in assembly one of the eggs hatched. The very next day the other duckling hatched. We called them Edd and Samuel. We fed them boiled egg and chick crumbs. They grew very quickly and were soon running and jumping all over the place.

If you saw our assembly on the 27th of June, you will know that ten days later, Edd died. Samuel (has) is now living with a very caring hen.

Alison McKinley
Age 10
Steeple Aston School

Thank you my duck!

In memory of Edd the Duck, class 5's Duckling. Born on the 6th of June, 1990 Rest in Peace

An Order of Service for the funeral of Edd the Duck.

Edd the Ducks Funeral service

Thank-you for all coming to Edd the Ducks funeral service.

He had lots of good friends, who are hopefully are all here. Thank-you to Lucy, who was like a mother to Edd, and to Samuel his dearest friend and Mrs. Gardner for making it worthwhile.

We are so sad that he has gone, we hope he has a good time in heaven.

In loving memory of Edd the duck,
His life was so short but full of frien
And even when your so far away,
Our love for you will never end.

God please look after our duck,
As well as you possibly can,
Give him the very best of luck,
While we comfort his brother sam.

Please join together and sing kum ba
Burial
Hands together and eyes closed...

The children of Dr. Radcliffe's School, Steeple Aston held a funeral service for their dear Edd the duck, who had earlier hatched at school in one of the Sanctuary incubators.

Dear Lord,
Edd's life was very short,
But we know where his spirit is now,
That he will live in the very best of hands.
Please look after him for we really cared.

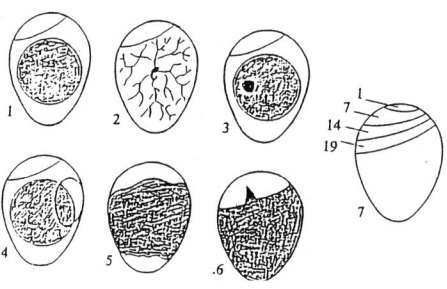

Egg Development (days)

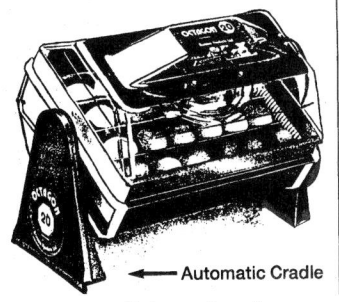

Brinsea Incubator ← Automatic Cradle

July 6th / 7th 1991

Saturday, Hook Norton Rural Fayre. We loaded up the tractor and trailer with large cages ready for an early start, but what with all the birds and animals to feed here first, it was hectic to say the least. Rod and Wendy were frantically trying to supply the feed all round as well as keeping the water bowls topped up. I myself rushed hither and thither in my usual crazy fashion — *"flapping around"* they call it. I prefer to call it being in charge and responsible for everything. We arrived at the show to find three cars parked in the shady patch allocated for our baby stock. There were three announcements on the tannoy requesting the last one to be removed, but with no response, they organised a tractor to tow it away. Meanwhile we were forced to set up our show in a sunny part, which meant we had to cover up quite a few of the birds from the sun. Our little patch proved extremely popular with the children, sitting on straw bales; it was full all day. Some children are quite terrified of holding baby animals. In one case, I succeeded in changing the approach of a particular little girl, by the end of the day. I took the opportunity of handing out hundreds of Sanctuary leaflets.

HOOK NORTON

The small village with the big heart

By ROSEMARY CLAYDON

HOOK Norton is a small village with a big heart and an even bigger community spirit, which is typified in the work of the Hook Norton Charitable Association.

Formed in 1987 to help the numerous unfunded village organisations in the village, the work of the Association has mushroomed beyond belief, and the list of its benefactors, both local and national, is endless.

Its initial impetus of small events culminated during that year in Hook Norton's first Rural Fair - a modest undertaking by today's standards, but successful. It became obvious that the Fair was destined to become the biggest money spinner and the attention of the committee was devoted to the preparations throughout the year for this one grand annual event.

And grand it has certainly become! Hook Norton Rural Fair has become well recognised as one of the outstanding events of the year in the Banbury area and far beyond, and its attractions have become increasingly sophisticated as the years have rolled by. From a gate counted in hundreds for that first event, last year's Show attracted almost 4,000 supporters, and the event was really and truly on the map!

The Show is run by a Steering Committee of just four people, headed by Roger Hughes, who has masterminded the event for the past two years and it is as a result of his inspiration and enthusiasm that the event has motivated the whole village, individuals and organisations from the Brownies to the Old People's Club, to give unstintingly and single-mindedly of their efforts and skills as the Fair approaches.

Virtually all village organisations and institutions have benefited over the years, from various clubs and societies to the local village hall and the impressive and successful Church Restoration Fund. Various outside and national causes have also been helped, main among which is the Leukaemia Research Fund in which the village has a special involvement.

Furry friend... six-year-old Alanna Winter gives a rabbit from the Wiggington Waterfowl Sanctuary pet's corner a hug at the Hook Norton Rural Fayre on Sunday. The organisers estimate that nearly 4,000 people attended the fayre, which raised over £7,000 for local charities.

July 8th 1991

A coach-load of schoolchildren from Middleton Cheney Infant school, around 50 altogether. All enjoyed themselves, and there were no mishaps. I took them round the Nature Trail, a lovely group of six-year-olds, full of life and enthusiasm, and we went close up to the Highland cattle, which have settled down well, and are quite unaffected by children. Some thought they were buffaloes, as the mature females have very long horns. Miss Griffiths, Mrs. Finch and Mrs. Smith provided a list of things to find to take back for a class project. The Sanctuary is certainly proving of great educational value.

July 9th 1991

Mothers with toddlers came in two coach-loads from Steventon Playgroup, nr. Abingdon, and stayed for four happy hours. In the evening I sat in my summer-house for half an hour and just felt so content looking out over the Sanctuary – all that I had created in the last few short years. What a beautiful evening it was, the sun setting in the west and everything here looking so relaxed and happy with life! A little baby wild or semi-wild rabbit came scampering out along the paths. The moorhen shyly picked her way across to the pond. Her eggs seem to have disappeared once more, no doubt taken by those wretched magpies, who are forever on the prowl, screeching and gliding over the whole area, spying out, with their oh so sharp eyes, any egg laid within view. The saddest aspect of the job that I do is noting the disappearance of precious eggs. Shortage of money this year is preventing me from buying electric eggs. I'll see if funds run to it next year. These are artificial eggs containing batteries, giving off a sharp 'buzz' when they are pecked.

July 10th 1991

Seventy-seven children came from Glory Farm School, Bicester, plus teachers and helpers, and all enjoyed their visit very much and many thanks were given. I had a nice compliment paid today, a couple from London said they preferred the Sanctuary for their little girl more than any other similar attraction.

July 13th 1991

My grandson Lee came with me to a pet store to buy feed for the birds etc. He tripped on a temporary paving slab in the shop, fell and cut his chin open. We dashed straight up to the Horton General Hospital where he had four stitches. Poor little man, he was very brave. We sped back to the store to buy the feed, then immediately out to the Sanctuary, where stocks were at a very low level.

Jeremy escorted a birthday party of fifteen little girls here this afternoon. He is going to spread the word about how the Sanctuary is such a perfect venue for birthday parties; there's no need for any other entertainment to be organised.

July 14th 1991

Lots of visitors today. Jacko the Jackdaw proved very amusing and quite popular with some of the men when he landed on their shoulders. He pecks at jewellery and buckles on people's sandals. With his long sharp beak he playfully tweeks at necks and faces, but doesn't hurt too much.

Late this afternoon, a Sunday, some folk from Swerford brought in a sack carrying a strange bird thinking it came from here, I had not seen one quite the same before. It was bleeding from an open wound in its belly and needed vet treatment, so I rang Kate, our lady Vet from West Bar Hospital who arranged to be at the Surgery within ten minutes, the time it takes me to get down there. First of all I gave the bird a half bucket of water with cod liver oil in it. He was delighted with that straightway getting into it, out and in again repeatedly. On inspection Kate thought that it had not been shot as the people had thought, since it was unable to fly, but it had torn its cut his underside on a very sharp edge. She gave it an injection of antibiotics immediately and hospitalised it overnight, giving me instructions to telephone tomorrow afternoon to see if it had survived. Rodney looked it up in his bird books and thought it looked like a glossy Ibis, the nearest to it. It was white with black feathers on the tail and wings, a head with a six-inch long beak, jet black and slightly curled under at the end; obviously it was a wading bird, but without webbed feet, its long thin, black legs resembling a stork or egret like the ones I saw in Israel.

July 16th 1991

Mr. Hughes brought children from Freeland School, combining a visit with a class project, and took them up to the Nature Trail. We all got caught in torrential rain after a lovely sunny morning. We sheltered beneath the observation tower but still got soaked through even so; the kids were all happy, and loved every minute; after lunch, the sun shone again; they did their questionnaires, set off back home at 2.30 after a very successful visit.

At 3 pm. the bird brought yesterday was still alive, and the vet rang to request me to bring some fish for it. Temporarily, I had to buy frozen cod. The vet had stitched up three tears; no internal damage showed up on the X-ray: only severe bruising. I popped to Hooky to pick up a cage offered by a well-wisher. On my return, a war-cry was sounding throughout the whole Sanctuary. I dashed through and saw a fox disappearing through the top gate. He had chased a white Angora rabbit, John One, around the lawn area, fur was scattered about, and he was very frightened, his heart thumping when I picked him up, but otherwise he was no worse for his brush with death. Wendy had just groomed him and trimmed his thick coat, giving him a run outside on the lawn, as we frequently do with our breeding stock of rabbits, to prevent them from becoming bored with their existence. In the morning I went to collect bread from Middle Aston: this should work out a little less expensive, helping to feed the birds and animals until our wheat is ready after the harvest.

July 17th 1991

Three coaches today, Southfield School, Brackley, Henry Box School, Witney and Kidlington Mums and Toddlers group, plus a few others. An altogether lovely, happy day: no mishaps.

July 18th 1991

The ibis died at the vets. Rain most of day with very few visitors. Wendy and Rod are both ill with a flu-type virus. I had to do all the feeding myself: a long, slow job. I transferred the Andalusian mother into Reception cages, then added ten incubated chicks to her. She took to them immediately, all except one brown silkie; and this one I promptly removed back to the heat lamp with more day-old incubated chicks.. The kiddie visitors love day-old chicks and ducklings. I have been forced to remove teenage ducklings and put them in nursery houses as, for three nights running, stoats have taken their toll once again, then last night they killed a full-grown Silver Appleyard duck. It's always the females that are taken, it seems.

July 20th 1991

Swalcliffe Fête found me a nice little corner once again, proving as popular as ever with the children. Snowy, my largest Angora was the centre of attention as usual; she just lies on the top of the straw bales to be admired by one and all; she really is very beautiful with her long, thick, pure white fur and lovely serene face, large pink eyes and white fluffy tufts on top of her ears; she hops around, sometimes just to show she really is alive; she measuring about thirty inches long when stretched out.

July 22nd 1991

Bob Morris and family brought to the Sanctuary five pedigree Cotswold sheep, four ewes and one ram. They were about to be slaughtered, but family pets just must not die unnecessarily. It

Thank you my duck!

The Award Winning Water Fowl Sanctuary & Rescue Centre
Wigginton Heath, Nr. Hook Norton,
Banbury, Oxfordshire OX15 4LB

Eggs & Incubation
Notes for Teachers

Price: £1.95

Contents

	Page
Why Study Eggs?	2
Obtaining Fertilised Eggs & Incubators	3
The Reproductive System	3
Fertilisation	3
The Structure of the Egg	4
Development Within the Egg	4
Artificial versus Natural Incubation	5
Preparation & Storage of Eggs for Incubation	6
The Incubation Procedure	7
Hatching	8
The First Days of Life	9
Commercial Egg Production	10
	11

The Award Winning Water Fowl Sanctuary & Rescue Centre
Wigginton Heath, Nr. Hook Norton,
Banbury, Oxfordshire OX15 4LB
☎ 0608 - 730252

No part of this booklet may be copied without the express permission of the publisher.

Why Study Eggs?

Hens, which begin life as a speck on the yolk of an egg, can be laying eggs of their own within six months. The speed of this reproductive process makes the hen, eggs and incubation valuable educational resources.

The National Curriculum requires that pupils should:

- know that animals need certain conditions to sustain life,
- understand how living things are looked after and be able to treat them with care and consideration,
- know that living things reproduce their own kind ...
- know that personal hygiene, food, exercise, rest ... are important ...
- know that the basic life processes ... are common to all ...
- be able to describe the main stages in the human life cycle ...
- understand the process of reproduction in mammals ...
- know about factors which contribute to good health and body maintenance ... including balanced diet ...
- understand malnutrition and the relationship between diet, exercise, health, fitness ...

The incubation of eggs and the rearing of poultry species provides excellent opportunities for the introduction of the study of reproductive processes in vertebrates in general and mammals in particular.

In addition, such exercises enable pupils of all ages to develop a degree of insight into the use of eggs and poultry as sources of food. For most pupils, the experience of handling fertilised eggs, watching them hatch and caring for the chicks is highly rewarding.

Eggs and poultry are important elements of the diet of many people, both being major suppliers of protein. It is important that the nutritional aspects of egg and poultry production are appreciated.

2

was lucky we had a spare pen and fresh grazing. They look a bit tatty, but will soon spruce up on our land. I feel very proud the way everything looks here, the good condition of the animals under our care. Wendy is taking over the spraying, worming, disinfecting, dipping, and shearing — the many things needed to keep them healthy and happy in their time with us. Bob also brought a trio of farmyard geese. I did think I would refuse these ordinary cross-breeds, and just stick to rarities, but as I've had a few hatch this year I weakened and took in these three. I just can't let them die; they look up at you with bright blue eyes and are so appealing.

July 23rd 1991

Today Ruth, a large, black-faced sheep, was brought to us. She was left in an orchard, her owner having died very suddenly. She was a pet lamb and halter-trained, and the neighbours couldn't bring themselves to send her to market. She had an ear infection requiring urgent veterinary treatment. I immediately sprayed her with fly repellent and put antiseptic into the ear to relieve the obvious pain she was in. Wendy trimmed her feet and wormed her.

Quite a few little visitors as they are breaking up for holidays. I also received a lovely package of thank you letters from children of Freeland School. *See samples from each school at the back of the book.*

July 24th 1991

We've managed to bale 90 bales of hay, putting them safely in the barn; the remainder is cut but we're getting rain each day; it looks ominous, but we're still hopeful.

Wendy has been dealing with the sheep, moving them to new pens: the regular procedure and rotation of three weeks in one pen then on to a new one, allowing the grazed areas three weeks to recover. This we believe will check worm infestation. All of our stock looks well and healthy; we are still hatching chicks and ducklings. A black East Indian duck came off her nest with four tiny babies; she had been sitting on seven.

One baby peacock has been incubated artificially; a tiny partridge chick brought in from Moreton-in-Marsh, only one day old, but very lively – a true survivor. It's in the brooder with other day-olds holding its own, only the size of a milk bottle top. A Golden Pheasant died in the night; it was trapped between two layers of wire. I think it was frightened by a stoat or weasel — no visible injuries.

A foreign family with eight children visited and one boy of eleven or twelve picked up a baby chick and threw it down on the ground. I went wild at him, told him off in no uncertain manner. The woman was expecting yet another child. She did not even apologise for her son's behaviour, so I said to her *"will he throw your baby on the floor when it's born?"* They all stood and looked and looked, but not one of them said *"Sorry"*; I kept a very watchful eye on all of them until they had left. Oh dear! It is not in my nature to be angry, but if someone hurts my babies, I blow my top. My experience is that English children are ninety-nine per cent kind to birds and animals. One little English girl, however, did kick a tiny cockerel who comes into the reception – no reason at all, as he is not a spiteful cockerel. I asked her *"Now did that little bird hurt you?"*, *"No"*. *"Then why did you kick him?"* She looked up at me not quite knowing whether to cry or not; however, I hope the reprimand was enough to teach her not to provoke animals, or human beings for that matter, into retaliation in anger. (I want to teach the world to sing in perfect harmony). What a wonderful place the world would be if everyone learnt always to be kind to each other when young. I shall persevere in my task of helping to sow the seeds of a small amount of kindness, by allowing children of all ages to cuddle my babies.

The small American boy called Evan, who came to England for a week's holiday to a nearby village, re-visited with his host on Monday, again on Wednesday and then brought his Mum along on Friday. He's really fallen in love with our place.

One Angora rabbit gave birth to nine and let them all die: no milk, no nest. Not a good mother. Best not to let her breed again; it hurts me to find something like this.

JULY 27TH 1991

A beautiful kestrel picked up from West Bar Veterinary Hospital, uninjured but just concussed. It will fly away again in a day or two. I also collected a thrush, with a broken wing, which the vet had taped up. This will need to convalesce longer before it can fly again. The little partridge chick is doing well, holding its own beautifully; the peacock chick, now six days old, likes to fly out of the brooder tank and comes across the room to be picked up; I carry him in my pocket and he nestles down quite happily, stopping his permanent *"cheep-cheep"*. Looking to the future I can hardly envisage a fully grown peacock trying to climb into my pocket! It conjures up quite a picture! He seems very lonely in spite of being raised with other young chicks and ducklings who are all quite contented. I'll name him Pointer the Peacock as he has such a poignant call.

JULY 29TH 1991

Either stoats, a weasels or rats have killed fifteen of my young rare breeds: one Poland, all three lavender Pekins, all black Pekins, all black Orpingtons, all gold and silver laced Wyandottes. The predators had got into the nursery house through a small hole in the tin sheets. I feel like weeping today. I have a whole year to wait for more. They produce so few fertile eggs, then something like this happens!

JULY 30TH 1991

Sold two young ducks, black East Indian, and one miniature silver Appleyard, £10 each.

Tim, a natural wildlife photographer from Chipping Norton visited, and he was very interested in Jacko the Jackdaw. He would like to include him on a film about jackdaws for the *Anglia Television "Survival"* series.

AUGUST 1ST 1991

Rodney is 21 today. He didn't want a party or any fuss made.

Andrew, a photographer, came from Bicester to write an article about the Sanctuary for a newspaper. Yet another unwanted rabbit from *Crofts*; it had started to attack the girls – frustrated at being in a small cage for too long, I guess. It's now in a big grassy run, and able to dig holes when it feels like it, just as most of the others do here.

Kate, a very pleasant and kind young lady from West Bar Veterinary Hospital, came specially to treat the sheep; she gave injections and advice. All staff there are so understanding and helpful.

The baby peacock came outside to find me, then two mature cock birds must have recognised it as one of their own species; they were both displaying to the best of their ability, as they have not yet become fully plumed. They were walking round and round this tiny mite. It was quite a funny sight to be sure, no larger than a blackbird.

AUGUST 3RD 1991

I took a Pets Corner to Newbold-on-Stour Village Fête in aid of their church.

AUGUST 4TH 1991

My darling baby peacock died – so sad!

AUGUST 5TH 1991

Jacko the Jackdaw takes shiny coins from our donation pot and drops them in the car park etc. He is quite a character. Rodney is very fond of him. He communicates with him, talking in his own language. If Rod is reading a newspaper Jacko tries to rip it up until Rod resumes paying attention to him. When he tells him he's not speaking to him and turns his head away the bird immediately hops round to the other shoulder so that he's being looked at again. What a character Jacko is! A few of the visitors are a little in awe of him, but so far he's not attacked anyone, (drawn blood I mean). He tries to peck shiny earrings, necklaces and buckles. He gathers up new shiny five pence pieces and stores them up on high ledges in the reception barn. Imagine in the future someone's surprise finding a stack of coins etc. high up there. It is a well-known fact that Jackdaws steal shiny jewellery from people's dressing tables when windows are left open. I had never seen such happenings until Jacko came. When a visitor asks *"Has anyone seen a bunch of keys?"* I immediately think, *"Oh dear! Where's Jacko?"*. One lady found hers in her own jeans pocket after searching around for sometime. I had begun to blame Jacko for the disappearance, but this time he was not guilty. He was left in reception overnight, and killed four baby chicks by picking them out of the display tank. Unfortunately I had not replaced the wire grill on top. My fault!

AUGUST 11TH 1991

Two dwarf lop-eared rabbits with cages arrived with a lady who was moving house. She cried a few tears at leaving her pets. She had cared for them for the last three years.

Many happy faces throughout the day. Roger the Angora goat escaped and followed visitors around the Sanctuary – quite harmless; he seemed to be a great attraction, roaming free, and sharing picnics at the picnic tables.

AUGUST 12TH 1991

A large Terrapin, recently picked up at Wigginton and brought back by an off-duty policeman, has settled in well.

AUGUST 13TH 1991

Tim Shepherd has been at the Sanctuary all day filming Jacko our Jackdaw for the *"Survival"* Series to be shown on ITV in the future. That will be interesting; I hope they mention the Sanctuary.

Rats or stoats killed my three baby black East Indian ducklings. I had put them out on the lawn in an ark with their mother. The creature buried underneath and took them through two-inch wire mesh.

Walter Went is building me four really tough houses for my babies, rat proof, stoat proof, and mink proof, with £200 donated to the Sanctuary by a very kind American lady, called Betty. She made only one condition: that, for one calendar year, I do not refuse entry to a family of little children whose parents quite obviously cannot afford the £1 admittance fee .

AUGUST 16TH 1991

A beautiful full-page centre spread in the Buckingham Review Newspaper, *(partly reproduced on page 86)*, — the best story anyone has written about the Sanctuary. Thank you.

W.I. from Garsington visited today all seemed happy with what they saw.

AUGUST 20TH 1991

Today Tim Shepherd was filming Jacko, including shots on Rodney's shoulder. He had endless patience waiting for the bird to co-operate — a day's work. What a pleasant young man Tim is: no arrogance, just gentleness and kindness towards birds, animals and humans alike.

Many visitors again as a result of the newspaper story. It has done more good than any paid advertising. Happy pictures, with accompanying editorial, are much more eloquent to readers than anything else.

A black ten-year old goat, called Claire, joined our ever-growing family. Her owner had suffered a stroke so he sent her to the Sanctuary in the happy knowledge of her safety and well-being for the remainder of her life. She will be joining others up on the sloping rough field at the top.

I have at last found a good home at Whichford for Thomas, the butting goat. He has gone to a family who will play rough and tumble with him and are not afraid of his horns.

Today I bought four peach-faced love-birds for £40 from a chap in Banbury, adding plenty of colour to the aviaries.

from the "Buckingham Advertiser & Review Series" August 16th 1991

Follow the duck to an animal kingdom where dreams come true

Mabel has 2,500 babies

MABEL Warner's family is bigger than the average... every day she has more than 2,500 hungry mouths to feed.

The 'family' is the award-winning water fowl sanctuary at Wigginton Heath, near Hook Norton.

The 22-acre sanctuary is home for waterfowl, chipmunks, guinea pigs, goats, sheep, lambs, rabbits, ponies, antic, goldfish, terrapins, exotic birds a pig and mischievous jackdaw.

The sanctuary is a dream come true for Mabel.

"It was an ambition of mine and my first husband's for a long time," she explained.

"The sanctuary is family-run and it's the happiness it creates that makes work self-satisfying."

Previously, Mabel was a florist for 25 years, but it was the childhood years she spent on her parents' farm that brought her closer to animals.

In 1987 she bought the land at Wigginton Heath and within two years the sanctuary was up and running and open to the public.

So far, thousands of visitors have been attracted to the wildfowl sanctuary.

But it is not only the wide variety of species people come to enjoy, some bring injured birds and animals in need of convalescence.

Mabel explained: "The land here has never been sprayed with herbicides or pesticides so when the injured birds are released they have all the natural food they could wish for in the grounds of the sanctuary."

Mabel also takes in animals that have been in pet shops for a long time.

"One owner asked me to take a rabbit because it was getting vicious, after a few weeks here it was as right as rain," she said.

All the animals are fed on herbs and stress tablets, a recipe which rids them of 'nasty streaks' says Mabel.

There is no other place where you can pick up all the ducklings, chicks and baby rabbits, she claims.

"We encourage this the whole time because it makes the animals tamer and it teaches children to be gentle with them."

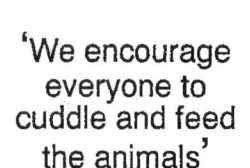

The sanctuary has a huge pen where children can sit down and feed rabbits. The pen is extremely popular with the public despite a warning sign at the entrance which reads 'Bunches Bite and Nibble, they are always hungry and quite like skin'.

Wigginton is home to a huge selection of wild and water fowl.

Pictures by Steve Nevard
Words by Andrew Laxton

There are Ruddy shellducks, Dutchcall ducks, Australian shellducks, Black and Bronze turkeys, Indian Blue peafowl, Black East Indian duck, Pink Footed geese, Buff Orpington hens, Cochin Partridge hens, Cackling Canada geese and Buffbacks to name but a few.

Walking around the pleasant grounds you will also come across a Sandy and Black pig, Angora goats, sheep, aviary birds, owls and a multitude of inquisitive yet friendly species.

There is an adventure playground for energetic children and plenty of places to sit for those with tired feet or to eat a picnic.

Stars of the centre are Snowy, an Angora rabbit, and Honeybunch, a Cashmere rabbit, who both lazily lounge on top of their cages and an owl which keeps a beady eye on all visitors as they enter.

Mabel sells some of her stock and has invented a new breed of bunnies called Mabel's Midget.

She breeds from rabbits before they are fully grown so the offspring stay small.

"There is nothing that can hurt the children here – everything is safely fenced off," she explained.

"Parents can come along and not worry about their children for a few hours."

Mabel likens the sanctuary to the tiny Swiss village of Pestalozzi, which is home to a number of different nationalities.

"It's exactly the same here, we have 2,500 animals and birds and they all live happily together," she said.

And judging by comments made in the visitors book the waterfowl sanctuary is the place to go this summer.

In 1990 the centre won the Oxfordshire Special Conservation Award.

Admission is £2 for adults and £1 for children, parking is free.

Opening times are from 10.30am until dusk all year round, apart from Christmas Day and the grounds are accessible for those using wheelchairs.

The sanctuary is very well signed from the A361 Banbury to Chipping Norton Road – just follow the ducks on the signposts.

For group bookings ring 0608 730252/677391.

'We encourage everyone to cuddle and feed the animals'

AUGUST 22ND 1991

A lady named Ali rang to say she had found a cygnet, (a young swan), in her garden. This poor little "ugly duckling" must have walked as it could not yet fly. It had still not feathered up, just plain, grey down; no obvious injury as far as I could see. Also, later, a white fantail pigeon arrived, apparently shot on the underside of its wing; this should recover well.

The response to the editorial story and photos continues to be quite astonishing.

A young lad, Ben, from Bodicote, was moving north with his family and had to part with his beloved ducks. He brought them to the Sanctuary and visited them on the evening of their departure and didn't really want to leave. It does seem sad that a boy has to be torn apart from his pets in this way.

AUGUST 23RD 1991

Yet another "ugly duckling" (cygnet), brought in by RSPCA, was found lost and disorientated at Preston Bisset. It is slightly lame, but eating well and looks quite bright. One will never know what happened to the parents of these strays; possibly they have been killed or badly injured. It's a pity the Swan Watch has been forced to cease its activities due to lack of voluntary support. The Swan Sanctuary down south said they would take the babies but I am happy to keep them here until they are old enough to fly away.

AUGUST 26TH 1991

This has been a good Bank Holiday Monday, with many pleasant visitors. One lady commented how everybody seemed so nice here and no litter around.

AUGUST 27TH 1991

In the last few days Den and Rodney have been cleaning out the barn at Church Enstone, lining the floor and walls with thick black plastic ready to receive the wheat which Mr. Cherry is combining for us. Hopefully, this year it will keep the stock fed through until next year's harvest, enabling us to save considerable expenditure. This pleased my bank manager who seems to be 'squeezing' me a little for the return of borrowed monies. Thank God for good weather and plenty of visitors each day!

A little girl from Tadmarton said to me: *"I know I shouldn't talk about you being old or dying, but when you die can I have all of this please? I do love it here"*, (meaning my Sanctuary). I answered: *"If you marry my son you would have it and nothing would make me any happier for someone to love it as much as I do."* The little girl is only ten and Rodney, 21. Time will tell!

AUGUST 29TH 1991

A tame crow named Jimbo brought in this morning. A family could no longer keep him as he had outgrown his cage. He is quite a character, similar to Jacko.

AUGUST 30TH 1991

Wendy Humphries, Mayor of Banbury, kindly sent photos to a few newspapers, with a story about the Sanctuary and with a plea for support.

Olitwo

August 31st 1991

Olitwo, the tawny owl, comes out each day and just sits on his log. Visitors love him, and are quite fascinated with his big eyes staring and blinking and by the double eye-lids. The inner membrane closes when owls in the wild dive for their food, yet they can still see through them. The outer ones are only for sleep; also people are very intrigued to see the the head turning around completely.

September 3rd 1991

We are still hatching baby chicks, one or two daily, with mother hens adopting most of them. A little sitting call-duck hatched four Poland bantams and also other hen eggs. She is still sitting and doesn't come off her nest with the baby chicks, but just sits tight — a truly wonderful little foster mum.

Some pigeons have been taking it in turn to foster three baby black rabbits, who were born in nests on the floor of their aviaries. Now they are wondering what's happening to "their babies" as these are now ten/eleven days old and are beginning to leave the nests from beneath them. Normally baby pigeons, known as squabs, would grow to full-size before attempting to fly. This is why one seldom sees fledgling pigeons. Melody photographed the rabbits with one mother pigeon yesterday. They have grown double the size of the ones born on the same day to another doe in another corner of the aviary. This is because of the extra warmth given from the pigeons' bodies, I assume. Pigeons fly above, rabbits live safely and happily below.

Two young pigeons brought in from Banbury. Builders had erected scaffolding around some offices and destroyed the parents' nest site. The young birds had hopped all the way up the stairs and the kind-hearted office girls had brought them to me.

Stoneleigh 1991

September 7th 1991

I spent all day at Stoneleigh, Warwickshire, today, stewarding for the British Waterfowl Association — something I had never done before. I enjoyed meeting Mike Hatcher, Roy Price and Roz Taylor after reading so much about them and the hard work they put into running the Association. Tom Bartlett was as usual very busy at the show. He is the leading breeder in the country and a top judge. Along with Michael Roberts of the *Domestic Fowl Trust*, these two have, I believe, contributed more than anyone else to the breeding of rare poultry in the British Isles.

Tony Axton from Devon won the B.W.A. Cup for top marks, presented by Tom Bartlett. Prices at the sale seemed quite low for some species. Some were withdrawn by their owners because they did not reach the reserve price, whilst others made very high prices, e.g. £230 for best white call-ducks, a perfect pair, top prize-winners in the show.

September 13th 1991

Mrs. Mead from Lois Weedon, Northants., brought in a pair of Asian magpies, a beautiful blue in colour. She had kept them for eight years. I hope I can go on keeping them alive and well.

September 14th 1991

Bridget from Hook Norton brought along *Ping Pong*, a white duck, with a crest on its head — a breed known as the Crested Duck — from one of the eggs their little bantam had hatched earlier in the season. They also brought an Appenzella Spitzhauben bantam called *Grace Jones*; another, a Pekin called *Spats* because of the feathers growing on its feet and legs. Bridget thinks up the most amazing names for her family of poultry! Being a musician, however, she cannot give them the attention she felt they deserved – often going up to London for days on end. Her poultry pets are safer here.

JANUARY 12TH 1992

Steven, a little boy from Bladon received a five pound note for Christmas. He told his mother he was going to adopt a duck at the Waterfowl Sanctuary to help feed all the animals and birds. There's such hope for the world when little boys like that are growing up with such a caring nature! Maybe one day Steven will own a place like mine.

JANUARY 13TH 1992

A well-wisher sent a £300 cheque to help feed the animals and birds through the winter. God bless you, Jane, it will help a lot.

I treated Penny with *Spot-on* antiseptic as she has a small patch of scale. Everyone loves friendly Penny, the Shetland pony.

A fox killed one of the male peacocks last night. He had roosted up on top of the aviary where I had caged up his wife to give her treatment for an ailment, as she was beginning to look a little sick. By evening she had quite recovered but has sadly lost her devoted husband. If only he had stayed roosting up in the oak tree like all the others he would still be alive today. If any animal strays outside the electric fencing at night they become fox food, I'm afraid!

JANUARY 18TH 1992

First of the baby lambs has been born to the Cotswold mother who came to us last summer needing a lot of treatment. Wendy did all the work on her to make her into the beautiful sheep she is today. Her fleece looks shiny and silky in lovely condition, and her lamb is strong and delightful.

Banbury Cake came and took photos of masses of baby rabbits in a tiered basket set up by photographer, Georgina. The recession is hitting even them, as people are advertising less these days.

My old friends Jim and Doris Pegram (from Chinnor), their son Michael and Jeanette, his wife, came to visit — it was lovely to see friends again from forty years odd years ago. After long and friendly chatting, Doris said she could stand here forever and listen to the stories about the Sanctuary!

A little bantam is sitting on nine Rhode Island Red eggs.

JANUARY 23RD 1992

A second lamb born to one of the Cotswold sheep.

JANUARY 29TH 1992

One lady brought in five young guinea pigs and another brought a pigeon she had kept indoors for three weeks. This flew straight out of the box and joined the rest of the family – now making approximately 150 here, flying free.

Poppy and the Golden Guernsey Goats

JANUARY 30TH 1992

Sadly, Poppy, our dear Golden Guernsey goat died. *(See colour plate)*. She was very old when Jane first brought her to the Sanctuary nearly five years ago before we opened to the public. She had a very happy life with us in the top fields, with plenty of space. She was never bullied, her coat shone, and she was much photographed by admiring visitors. She just went to sleep peacefully in her old age without apparent pain. Wendy kept her feet clipped regularly to keep her comfortable.

I was told of one very interesting story concerning Golden Guernsey goats. In the Second World War when the Germans invaded Guernsey and were killing every goat they could lay their hands on for food, a certain old lady took her herd of Golden Guernsey goats deep into hidden caves and lived with them throughout the remainder of the War. She and her goats had food smuggled to them by well-wishers and friends, who risked their lives to do so. This is how she preserved the extremely rare Golden Guernsey goat for future generations to enjoy.

FEBRUARY 1992

The education pack is now ready after much work preparing drawings for worksheets and booklets for teachers. The design is by Derrick Butcher of "Main Stream", Cumbria, an educational specialist in this field matching material to a place as unique as Wigginton. He has produced twelve two-sided fact sheets; also included is a twenty-six page book on incubation and hatching, descriptive leaflets telling visitors about the Sanctuary, maps, headed paper, all with a specially designed logo featuring three runner ducks. The whole package cost approximately £2,500. I shall soon be sending out booking forms, order forms, leaflets etc. to surrounding schools.

The *Banbury Guardian* featured the launch of our education pack, thanks to Rosemary.

MARCH 3RD 1992

Wendy's pet lamb of last year, Larryadne, gave birth to twin lambs and took to one but rejected the other. It will be a bottle-fed friend for the little Jacob which has been brought in, one of triplets. Melody named the Jacob lamb Sarah; she is a survivor and growing well. Wendy's friend, Jeff, gave her a tiny Texel lamb weighing only about one and half pounds. He thought it would die, but it is gaining weight well. Wendy now has three pet lambs to look after and feeds them on powdered milk from a bottle about every two hours.

MARCH 16TH 1992

We have many baby rabbits now, they're as popular as ever; quite a few chicks are hatching. Three bantams were born with deformities, unable to stand up. This sometimes happens with the rare breeds, especially when their history is unknown. It's possible that inbreeding is the cause; something I constantly try to avoid.

Michael Page is hatching Marans, Wellsummers and Rhode Island Reds and I am taking them on from one day old. We are experimenting with offspring of my very high quality cockerels, who appear to produce a good quantity of females as opposed to males. High percentages of cockerels are of no use to anyone. My super Maran cockerel died, but Michael and I have three of his sons. Maybe in time they will prove to be as good as their father.

I have received this heart-warming letter from the First Sibford Brownies. I am very grateful that the brownies themselves have chosen to support the Sanctuary.

God bless them all!

MARCH 25TH 1992

Little Billy, the young male goat, had obviously worked well as one after another, within three days, three sets of twin kids were born to the white nannies. We have four pet lambs now this year whose mothers couldn't cope with

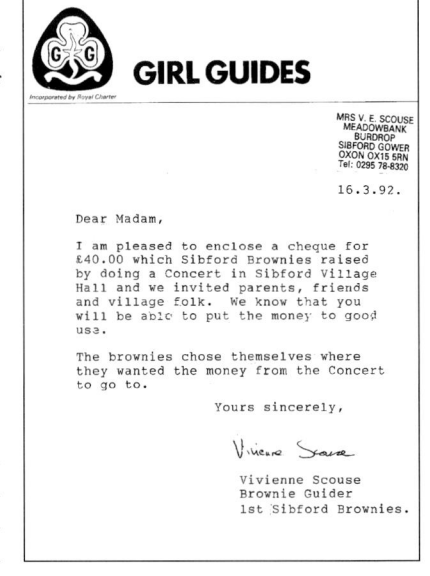

Thank you my duck!

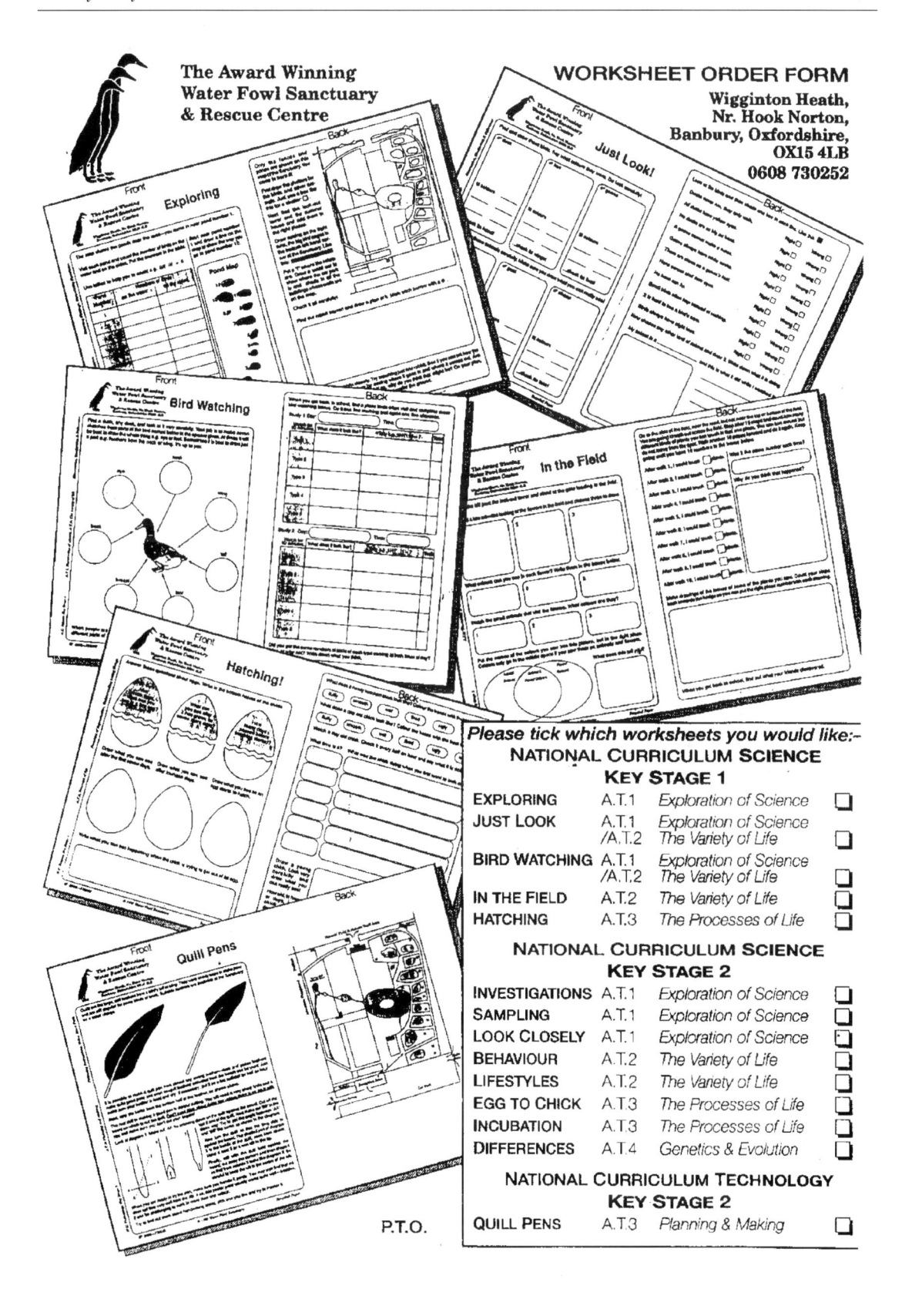

The Award Winning Water Fowl Sanctuary & Rescue Centre

WORKSHEET ORDER FORM
Wigginton Heath,
Nr. Hook Norton,
Banbury, Oxfordshire,
OX15 4LB
0608 730252

Please tick which worksheets you would like:-

NATIONAL CURRICULUM SCIENCE
KEY STAGE 1

EXPLORING	A.T.1	*Exploration of Science* ☐
JUST LOOK	A.T.1 /A.T.2	*Exploration of Science The Variety of Life* ☐
BIRD WATCHING	A.T.1 /A.T.2	*Exploration of Science The Variety of Life* ☐
IN THE FIELD	A.T.2	*The Variety of Life* ☐
HATCHING	A.T.3	*The Processes of Life* ☐

NATIONAL CURRICULUM SCIENCE
KEY STAGE 2

INVESTIGATIONS	A.T.1	*Exploration of Science* ☐
SAMPLING	A.T.1	*Exploration of Science* ☐
LOOK CLOSELY	A.T.1	*Exploration of Science* ☐
BEHAVIOUR	A.T.2	*The Variety of Life* ☐
LIFESTYLES	A.T.2	*The Variety of Life* ☐
EGG TO CHICK	A.T.3	*The Processes of Life* ☐
INCUBATION	A.T.3	*The Processes of Life* ☐
DIFFERENCES	A.T.4	*Genetics & Evolution* ☐

NATIONAL CURRICULUM TECHNOLOGY
KEY STAGE 2

QUILL PENS	A.T.3	*Planning & Making* ☐

P.T.O.

them for one reason or another. Chicks are hatching daily; I sold my first two Sebastopol goslings, the first this year. All three, grandmother, mother and daughter are sitting on nests of eggs. The Brecons are laying well, also the grey and white Chinese, but I am selling most of the farmyard goose eggs for decorating: exquisite creations – some containing little doors and windows are made by clever artistic people. The blue Asian magpies are showing an interest in the nesting material I provided, including a large basket. I do not know how they will tell the difference between their own eggs, (very valuable), and the bantam eggs. I have to feed them daily. I suppose nature tells them not to eat their own, otherwise there would be no offspring. The wild magpies are my worst enemy where eggs are concerned. They hover around the Sanctuary every day looking for any egg in sight, and will even go into the shelters and steal them. They even carry goose eggs to the top field.

I now have at least one hundred and fifty goldfish in the pond. They give me so much pleasure. Just standing and watching them for a few minutes is such good therapy. Frog spawn has appeared in all of the top ponds in the natural area where ducks do not go, so our family of frogs is increasing. I haven't spotted any newts yet, but no doubt they are around. Although I introduced several toads last year, I am reliably told frogs and toads do not live together in the same area.

Surrogate Mothers

MARCH 26TH 1992

One little kid goat is not well. Wendy feeds it on the bottle, gives antibiotics, puts it on an electric blanket to warm it up again. She thought at one point it had died, but it responded well through the day, even once standing up. The morning it was born, (one of twins), Wendy spotted the mother goat with one little kid at her side and the other, a short distance away, being protected by Penny, our Shetland pony. The pony seemed to be taking the kid goat for her own, neither letting the mother goat near it nor letting the kid suckle its mother. This was serious, for if the kid did not get the first colostrum fluid a mother produces for her first-born, it was likely to die as a result of the deprivation of antibodies. When we went up to bring the little family into shelter, (it was snowing at the time), Penny defied us to take *her* baby away, galloping and prancing round and round the little mite, and standing right over the top of it as if to say *"it's mine, it's mine, she's got one, I want the other one."* She dashed around the field when we picked it up, its mother following slowly into the shelter. Penny made such a song and dance about it, neighing, tossing her head, her whole body a-quiver with emotion. It was quite an extraordinary experience. The little goat survived and has been named Friendly Fred.

With Wendy's care, Tex, the orphan lamb, was also saved. To complete this particular trio she cared for an Aylesbury duckling which was brought by a lady from Enstone who keeps *The Swan* guest house. Her son, on a trip to a farm near Witney to buy ducks for catering, saw this tiny mite and took it home to give his mum as a Mother's Day present. Anyone who has experience of ducklings will know they are lovely but very messy, especially underneath dining tables attended by paying guests. She very reluctantly brought the baby, (named Marianne, an alternative name for 'mother'), to me at the Rescue Centre, where I placed it alongside the soft cuddly warm bodies of the kid and the lamb who sleep on an electric baby pig blanket. She snuggled up to her adopted brothers, and now the trio are inseparable and go everywhere together. When Wendy goes up to the top pens to tend the other sheep and goats she is followed by Tex, Fred and Marianne, all in a line like shadows and making a very pretty picture.

APRIL 5TH 1992

Sunday, a glorious day with lots of visitors. Mr. Potter from Bloxham brought Alex and Melanie, Amy and Caroline — (all were from Bloxham: Alex Potter of Butler Close; Melanie of Chipperfield Park Road; and both Amy Ratcliffe and Caroline Broughall from Greenhill Park). — They handed

Tex, Fred and Marianne

me £48 which they had collected by sponsorship doing a twenty four hour fast to raise money for the animals and birds here – wonderfully kind little girls. I know full well how much children love eating, so it must have been especially hard. I thank the Mums, Dads and helpers who encouraged these four very special little girls. *See colour plate.*

I sold five of my young doves, £5 each, also three of my own breed baby rabbits, the prettiest little fluffy ones, Cashmere crosses, small bodies, quite adorable for £10 each.

APRIL 20TH 1992

A wonderful weekend for Easter with a record number of visitors, everyone happy – a good omen for the future. Fred, the little friendly kid and Marianne the duckling and Tex are favourites with most visitors. A lady in a wheelchair cuddled the baby kid. *See colour plate.*

APRIL 21ST 1992

Easter Tuesday, quite busy.

APRIL 22ND 1992

A coach-load of children and helpers came from Stoke Mandeville hospital through the Aylesbury Area Health Authority. All enjoyed their visit. One helper said it was the first day for many years that one particular little girl was not whimpering, so it must have been good therapy for them all.

APRIL 23RD 1992

Very pleasant visitors again. One gentleman asked: *"could you tell me why every one of your 2,000 odd animals and birds look so happy and contented living communally, sharing pens and houses?"* My answer to him was *"LOVE"*. I love them all, and they know it. It is a kind of telepathy between me and my family of birds and beasts. I supply enough feed in all their different environments, a puggling place for ducks, as they are never happier than when they are puggling along in the mud. They find lots of their feed at the bottom of the streams and ponds.

APRIL 25TH 1992

My kind American friend and benefactor, Betty Hanff, gave me another £200 to help care for the birds and animals. It will buy two Brinsea incubators to encourage schools to incubate eggs in the classroom.

APRIL 28TH 1992

The pair of blue magpies had laid seven eggs due to hatch on Tuesday. Great disappointment when I checked them: there was only one left. The others had all disappeared. I presume the female had eaten them, so if and when she lays another clutch I shall remove them and try to incubate them artificially.

MAY 9TH 1992

Another batch of five blue magpie eggs now laid, I have been putting them into the incubator daily. She's sitting on the one left in the nest. I hope to re-introduce the others when this one is hatched.

First baby budgie in nest-box. A lady and gentleman from Stratford brought five green budgies to add to the aviary. My friends and advisers, John and Margaret Radway, brought five grey cockatiels for me to sell to raise funds, (kind people). They are experts on rearing small parrot-like birds in their home. They have a Parrot Room.

MAY 16TH 1992

Michael Page hatched some more bantam eggs, one Poland, with a perfect white topknot. The silver cock brought by a visitor from Kensington, London, has mated successfully with my small black & white. I'm pleased about that. We've had many visitors this lovely sunny Sunday – it should please the bank manager.

A nest of baby robins was brought in; cats had killed the parents. They were dehydrated and very cold in spite of the heat lamp and being fed with a dropper. I could not manage to save these particular ones.

1st Cropredy Brownies raised £35 for Sanctuary funds, bless them.

A Tortoise called Fred

I boarded a tortoise called Fred, aged twenty years. He was the biggest worry of all during his four days here. The family who owned him loved him very much, stipulating they wanted him put outside in the daytime and put back in at night. The first night I forgot to take him home and had to rush back to the Sanctuary to pick him up at 10.30pm, worried that something awful would happen to him. Never again would I board a 'Fred', at least until I had good protection for such a precious animal.

MAY 17TH 1992

Michael Page brought twenty wee bantam chicks, — all doing well so far. Hatching is better now the weather is so warm. A few visitors daily. Sold my buff Cochins chicks for £3 each; also a blue for £4, two days old. The word is spreading that I sell young chicks of pure rare breeds. They are so lovely with their trousered legs and fluffy feathered feet.

MAY 18TH 1992

A little girl called Lucy from York, on her return visit, brought me a picture she had drawn of the

Sanctuary and wrote with love from Lucy. These lovely incidents make my life complete. Her daddy bought two little rabbits for her and two ducklings to take back home,

Best Place in the World

May 22nd 1992

Middle Barton school came using my designer fact sheets. The teacher said they all enjoyed it, and would have liked to have spent longer. They brought back the baby chicks hatched in their classroom – another of my education incubation classes. One little girl wrote in the Visitors' Book *"this is the best place in the world"* after being allowed to hold baby rabbits, ducks, chicks, mice and goslings.

May 23rd 1992

Thunderstorms in the afternoon. One dear old lady visitor was so terrified she shook like a leaf, so Wendy made her a cuppa. Micah and Melody, my grandchildren, and Wendy loved every minute of the thunder and lightning. The animals and birds were unperturbed.

Happy visitor Matthew Prince

May 25th 1992

Bank Holiday Monday. The first white peacock egg has been laid! I'll put it into the Incubator tomorrow, along with two blues.

A sparrow and a black crow have been brought in; also a racing pigeon, which had lost its way in yesterday's storms. Its wing number was the old 01 number of London. Having succeeded, after many hours, in tracing its owner, he callously told me that if I returned the bird, he would promptly wring its neck. The pigeon can stay here with the other hundred and fifty, and when he's rested and had enough food and water he will be free to fly away.

A baby mallard duckling, no more than 48 hours old, was brought from Wallingford by two caring ladies who had spent the last two days watching it paddle up and down the river looking for its mother and being rejected by the other groups on the same stretch of the Thames. The ladies told me they made clucking sounds to mimic a mother duck and the little lone one followed them home. They didn't know what to feed it on and so it came to join two tiny white runner ducklings who were hatched this morning. Straightaway it settled happily beneath a heat lamp in a warm brooder.

A Little Peruvian Giggle

May 28th 1992

Rain all day. Despite the weather a lady brought three children, one a little Peruvian Indian boy, called Daniel Beasley, about eight years old. He really jumped for joy with all the little baby things. The mice ran along his arm and over his shoulder and he chuckled a deep-throated joyous giggle which intrigued me. Oh, how I wished I had a tape recorder to record that lovely happy sound.

Another young couple came and said that they bought a little baby bunny from me a year ago and named it Mabel. They then read a newspaper article about the Sanctuary and realised they had named it after me!

A young man leaned on the Reception counter and gave me a lecture. He demanded to know why some animals were in cages. I asked if he would prefer them running loose as prey for predators? He said *"no-one should have pets"*. I replied that here they are given food, care and love and that little children find great comfort in having a pet of their own they can talk to them and share all their troubles with them.

COUNTRYSIDE EVENTS in CHERWELL 1992

20. SATURDAY 30TH MAY 14.00
SPRING HAS SPRUNG AT ADDERBURY LAKES
Enjoy the changing seasons at Adderbury Lakes Nature Reserve, a fascinating spot full of surprises. Accessible to wheelchair users. (See also event no 72). ADVANCE BOOKING ONLY.

21. SUNDAY 31ST MAY 14.30 – 16.30
WATERFOWL WONDERS
A chance to visit the award winning waterfowl sanctuary at Wigginton Heath. Come and visit our feathered friends and much much more. Those with walking disabilities are especially welcome as the site is wheelchair accessible. Entry fee charged but free tickets for the first 25 callers. Contact Mabel Warner on (0608) 730252.

22. SUNDAY 31ST MAY 14.30
OLD KIDLINGTON VILLAGE
Historical walk and talk around the interesting and unexpected in Old Kidlington.
START: Kidlington Baptist Church, High St/Moors Junction.
DURATION: 1/2 miles, 2 hours. GUIDE Gerald Gracey-Cox, Kidlington & District Historical Society.

Mayor Wendy Humphries & Melody

Sue Marchant & Olitwo

June 6th 1992

A lady who telephoned from Stockport last week arranged to bring her rabbit here today, as she is leaving the country for five years. She gave £50. Another couple, also going abroad, came with three ducks, (their pets for three years), and gave £100 for Sanctuary funds. They all love the atmosphere here, 'unpretentious' they call it.

June 1992

Four days running, bus-loads of American children came from Croughton Middle School, a little over-excited, but all very welcome. Betty brought £100 from a fund raising event for us.

N.E.C., Birmingham

Following an advertisement in *The Times Educational Supplement* I was invited to attend a June seminar and exhibition at the National Exhibition Centre in Birmingham at £180 for a stand instead of £400. Anthony agreed to accompany me. We took his TV and video, computer and slide projector. One of the three videos was taken by Mr. Walker from Tackley showing children cuddling the animals. Mr. Walker visited with his wife, (headmistress of Tackley Primary school), to loan me the tape. At the exhibition, interest was shown in the education pack and about fifty were taken. Hopefully, we will have resultant visits from the Birmingham area. The organisers awarded me a runner-up prize of a framed photo of my stand and an entry and editorial in their own brochure. They said ours was the stand attracting most comments and attention. It was the little baby chicks in the incubator that did the trick; a few people thought sparrows had got into the huge building when they heard the *"cheep, cheep"*, and others came from various floors just to look and say, *"Aaah!"* How wonderful that tiny day-old chicks can cause such a stir with adults as well as children! A large catalogue firm invited me to take part in distributing the incubators into the schools, supplying the eggs, *Brinsea* incubators and then taking the hatched birds back into the Sanctuary after a few days, calling it the Incubation Pack. (Not possible due to lack of suitable storage space.)

June 13th /14th 1992

Whilst away, a farmer's dogs from Wigginton got into the Sanctuary and killed my best little breeding buck rabbit, a miniature worth £300 (to me). They were seen by a near neighbour dragging their prey under my front gates. He tried to stop them but was in turn attacked by the two Jack Russells. On reporting the matter to the Police, I was told there was not enough evidence to prosecute. Early next morning I found a mother and her six babies had also been ripped from their cages; these were worth another £150.

June 20th/21st 1992

The barn owl hatched and was returned to Marcus in Banbury. Next day, a second one hatched; he is rearing them in his airing cupboard. I went to Shipston Rugby Club Rural Fair. I will not go again as it is not worth the trauma for the babies and animals. It is the wrong atmosphere – too much noise and smoke.

I received a package of thank you letters from the children of Neithrop School, Banbury, following their visit.

June 25th 1992

A lovely day and lots of happy visitors – only one miserable couple who grunted at the £2 each entrance fee, and then turned round and went out — (their loss). They just don't know what they are missing by not coming into the Sanctuary. Atmosphere cannot be purchased, it is created by happy animals and birds in peaceful surroundings, living naturally with plenty of freedom.

I had a good chat with Jean Price, (now a grandmother like me), and her husband, David, reliving old memories of the pets we used to keep and exchange and sell, e.g. for sixpence a rabbit. Jean and I were school friends and teenagers together from three years old. We reminisced on the time we missed seeing our boy friends, one of whom is now my second husband, Den. Due to a notoriously unpunctual country bus service and a vehicle breakdown, causing an unexplained cancelled date (in 1948). Jean and I were at the back of the cinema and saw the lads go down to the front. Not considering it good manners to disturb other cinema-goers in our row, we could not go forward to join them. Then we lost them completely at the end in the crowds, and we never saw them again. I thought the romance with Den was off and married someone else, Bill, just over two years later. Den and I met up again after I was widowed in the mid 1980s. Happily for me, he had never married. Den then told me he had written a letter saying how he wanted to see me again, (we were both then seventeen), but he had addressed it to me at the right farm but the wrong village! Ah well, it is no use looking back! I am so thankful I am now married to my first love.

Sue Taylor brought encapsulated maps, again helping to raise money for the Sanctuary. Sue brought a newspaper insert about a similar place to mine opening on July 4th at Wardington, the other side of Banbury. No doubt it will take away a few of our visitors. It is mainly word of mouth that brings my visitors on return visits, together with friends from many parts of the world.

June 27th 1992

Today I went to Deddington School Fête with my animals and enjoyed it there — less noise, and a lovely atmosphere with very happy children and parents. On my return I discovered that someone had broken the foot of one of my tiny rabbits, poor little thing. I strapped it with a splint and hope that it knits together. Almost every time I go away from my Sanctuary something bad happens, little or large.

June 30th 1992

A very happy school group came from Oxford and another group of Brownies and Guides later on.

The head teacher from Little Compton School came and was most impressed; he's a good ambassador for the Sanctuary. His school has chosen to come in the near future as he said there is so much to be learned here. I showed him the second baby barn owl incubated by Michael Page. The Guides and Brownies admired it before it was taken to Banbury to be reared by Marcus and Jane.

July 4th 1992

I hatched my first baby white peacock, followed by two more. My jammed incubator has been repaired by Geoffrey Pearce of Brinsea Incubators, my supplier.

July 9th 1992

I found a nest of a hundred and three guinea fowl eggs laid by just three who fly and roam free around the Sanctuary. One female is trying to hatch some of them, quite unsuccessfully, of course. They are laid in a deep, wedge-shaped nesting box, quite safe from any predators such as rats, magpies and crows. The egg shells are very hard to break, hence the large amount surviving. Guinea fowl are cruel to other birds and I had not planned to keep any more. If they take a dislike to others, they bully and peck quite viciously, even killing others larger than themselves. Their feather formation of black and white or silver grey/white is spectacular and set very delicately into the body. It is so easy when one catches and picks up a guinea fowl to remove feathers by mishandling the wrong part of the bird. A whole cloud of fluffy white down is released — so very different to the toughness of the eggs.

July 18th 1992

Many happy visitors came including Evan, a young lad from America. Last year he came again and again and has now once more returned on the first day of his holidays in England. What a special young man he is.

The small chairs are proving very successful they look so welcoming for toddlers to sit down and cuddle baby bunnies and chicks. They were purchased through a contact at the N.E.C. Managing Schools Exhibition.

George Johnson came again to do more video recording. He has been a very popular disc jockey at the Horton General Hospital for many years: voluntary work which has been much appreciated, supported by his dear wife, Ruby.

Michael Page has been quietly helping to hatch chicks for me. He is a good friend, gathering up eggs, marking them and putting them in his incubator. I pay him a little to cover electricity costs. The cuckoo Maran flock we are working on will hopefully produce the high percentage of female offspring we are expecting. Den bought the cockerel (1989) who produced very few sons, we are now using his few descendants for our breeding programme and are also working on a strain of Rhode Island Reds and Wellsummers. Thus far in 1992 the Marans are the most successful.

A man brought in fourteen speckled Belgian bantam cockerels, very attractive birds but quite useless, unless anyone happens to be looking for a pretty little cockerel. Most neighbours do not like cocks crowing. There are, of course, a few exceptions.

July 22nd 1992

A birthday party of thirty small children plus adults totalling approximately sixty plus the usual trickle of visitors: all these 'enjoyed' a lovely sunny day after the torrential rain yesterday. One child sat down in the mud, — plenty of work for the washing machine!

I had an evening appointment with Andy (headmaster) and Paul, from Neithrop School, who were bringing birds from their poultry unit, for me to care for while a new building is being built. At 9 p.m. the trailer arrived and they ferried it up through the poultry runs and into the peacock/pheasant pens, using spare spaces for thirty Croad Langshans, Belgians etc. It was so muddy that at one stage Paul slipped down in it on his backside. Not so funny at the end of a long day! By 10.30 pm. the birds were all in separate runs and houses. Five geese were put in the two topmost pens where the dew pond is fed by spring water. It always runs crystal clear in that area until it reaches the duck ponds and then changes colour, dependent on the amount of rain. Colours range from muddy brown to greeny shades and sometimes looks quite blue with the reflection of the sky. What a particularly caring teacher is Paul. They are a very lucky school to have him on the staff.

I re-connected the electric fencing and locked up, and went home, quite exhausted.

Ruby and Pearl's baby dwarf lops brought in to me from Stonesfield.

I heard a little girl explain *"disgusting"* and when I asked the reason she answered it was the water puddle in the pathway, I offered a trowel to divert is but she said *"no"*. I then told the story of the Little Red Hen. The moral is that not many want the work, but everyone wants the bread!

July 23rd 1992

Friendly Fred, the little white goat, jumped up on a picnic table and tried to eat a little boy's cake. Most of the family laughed it off but grandma wasn't too happy! I promptly shut Fred in a run, behind a wire fence. Poor Fred! Poor Grandma!

July 24th 1992

Lots of lovely little children at the start of their summer holidays; let's hope we get six weeks of fine weather.

Thames and Chiltern Tourist Board are in financial trouble and are closing down.

Today Jackie and Anthony, my daughter and son-in-law with their children moved to Barford St. Michael. Yet more rabbits brought in today; I sold my older baby fluffy lop for £15.

July 26th 1992

Wendy is 32 today.

A lady visitor said she was going to write about us in complimentary terms to *Blue Peter*, of BBC television. We can but hope.

The two guinea pigs brought in yesterday seem quarrelsome to other animals and so the baby goslings sharing the cage retaliated.

JULY 27TH 1992

A green woodpecker was brought in after being attacked by cats in Adderbury. It was not too badly injured as far as I can see and hope he is able to fly again, he had lovely colours on his head and body. He escaped by himself to freedom, straight out of the door of the aviary.

JULY 28TH 1992

A big white Angora rabbit has been brought in because he was born with blue eyes, and not pink eyes; therefore he's of no use on the show benches. He is so beautiful yet, according to the owners, he would have had to be put down had he not been able to come here.

JULY 29TH 1992

Little Fizzy, the five-week-old yellow gosling from my white Chinese geese, follows visitors all round the Sanctuary until he flops down exhausted. Some kind people carry him back down to the Reception Barn. He is very fond of people and they of him.

Taking a gander!

LIFE'S a hoot for Fizzogg the gosling. The five-week-old yellow Chinese gosling has taken to following visitors around his home at the Wigginton Waterfowl Sanctuary near Hook Norton.

Fluffy Fizzogg is not the sanctuary's only new recruit. It is also home to a group of lively five-week-old rabbits, nicknamed Mabel's Midgets.

Sanctuary owner Mabel Warner explained: "I call them Mabel's Midgets as they are a special breed of rabbit, bred at the sanctuary, that won't bite and won't scratch." if handled properly

The sanctuary is also currently housing Neithrop Infant and Junior School's collection of poultry while the school is building new homes for them.

"We are quite busy at the moment, and we've been having lots of visitors during the weekends," said Mabel.

Regular visitor Tiffany Leek, from Banbury, was happy to handle the animals for Cake photographer Steve Forrest, for the 15-year-old is hoping to become a vet.

Tiffany

Fizzy and friend

A pheasant brought in – injured by a car, he has lost his tail and has two large grazes on his body. When recovered he will be released.

AUGUST 2ND 1992

A little owl was brought in by Marcus Ridley so that he could be let free into the wild. However, on this very windy evening, he refused to fly away. I'll try again tomorrow as he would be a lot happier out on the wing, finding his own food in the top fields – there are plenty of mice and voles up there.

AUGUST 4TH 1992

The little owl is out of his cage, up in the beams. We've had to watch out for our baby chicks and tame mice and remember to put the wire tops back on.

AUGUST 5TH 1992

Lots of visitors today — there seem to be children bringing back their parents, brothers and sisters following on from their school visits. I let the children hear chicks hatching out, cheeping away inside the eggs, then later breaking out. They love it, the joy shows on their faces. It is thrill for them and even parents were asking if they could listen too, as they had never seen or heard this happening before.

Sometimes a few children are too excited after the cuddling session and do not read notices asking them to put the little babies back in the right cages. The babies are used to handling by me from birth every day of their lives, as were their parents and grandparents.

AUGUST 6TH 1992

A little girl and her brother, Eleanor and Edward Clarke, from The Gables, Churchill, Oxfordshire raised ten pounds for Sanctuary funds. Bless them! Other children want to help by raising sponsorship money, too. *See colour plate*

AUGUST 7TH 1992

A photographer has been sent by *The Times Educational Supplement*, London, for a photo session of children using the worksheets of the schools pack. The pictures are to accompany an editorial they plan to write soon.

AUGUST 12TH 1992

The little screech owl brought in last week refused to be released and kept coming back to Marcus. It is now living in the top rafters of the barn where we keep our baby rabbits, chicks and mice. Last night I forgot to cover up one container and the tame mice all disappeared. Who else could be the culprit except the little owl? He is nowhere to be seen this morning gone back to the wild with a full belly, I presume.

AUGUST 14TH 1992

A lovely little girl, Abigail, from London raised £37 for Sanctuary funds by a 'word bash' at her boarding school and came with her Dad to bring a cheque. Bless all who help the work here. *See colour plate*

This evening Den, Wendy and I were treated to a dinner by Anthony and Jackie as thank you for baby sitting Stephanie and Lee the day they moved house.

Helen, Ben, myself and friend

AUGUST 15TH 1992

A female ten-week-old Gloucester Old Spot piglet, was delivered to Sanctuary. She was exchanged for young point-of-lay hens and we now have to make a new pen for her as she is digging up the goat pen.

The Intimidating Bailiff

One of the very rare occasions of disappointment was that of a big butch type of man who bought two tiny rabbits in my absence. Wendy had a feeling we had not heard the last of him. Sure enough, at 9 a.m. today, two weeks later, he rang to say that one of the rabbits had died and he wanted his money back. At 5 p.m. he arrived and intimidated me by saying he was a bailiff, and continued threatening me until I gave him his money back. However, when I picked up the baby bunny, which he handed to me in a plastic bag, it was still warm. I draw my own conclusions to this sad affair: I believe it had just been killed by him, solely for the return of his money.

AUGUST 18TH 1992

I bought seven miniature Appleyard ducks, also a trio of buff-laced Wyandotte bantams for 1993 breeding.

Five Croad Langshan eggs have hatched from Paul's stock, — nice strong chicks. Children of all ages are, as usual, thrilled to see them emerging from their shells inside the all-round viewing Brinsea 20 incubator, which is especially manufactured for schools, being fully automatic and easy to use.

I'm disappointed with the article in the *Times Educational Supplement* (August 1992): this is not how I would have liked to have seen it presented. I was hoping to take it to Macclesfield House, the Oxfordshire Schools Headquarters, to help promote my Schools' Pack. No photographs were used despite the fact the paper sent along a photographer, who took a lot of time and patience with the children. Still, despite the expense of the teachers' packs the bank overdraft is coming down a little, so, in these recessionary times, perhaps I shouldn't grumble. The power of the pen is all important. Janette, the reporter, could not possibly see my Sanctuary through my eyes calling my Winniepiglet an unattractive-looking beached whale and great hoary monster! (She doesn't know the joy of raising a piglet from babyhood and the pleasurable grunts this half-ton Oxford Sandy and Black gives me in return for a rub along her back, thanking me with her loving eyes. Beauty is in the eye of the beholder. It is disappointing when writers do not feel the way I do about birds and animals, but it's hardly to be expected. They would have to be with them daily. I will try to write it all down, to the best of my ability. I was sent a copy of the page from the *Supplement* by Rachel, the young lady organising the next exhibition for managers and governors of Schools in London. Janette did, however, refer to me as a "Mother Theresa of animals"!

AUGUST 28TH 1992

Muriel, from Wantage, came to demonstrate spinning. (*See colour plate*)

THE BIGGEST THREAT OF ALL

AUGUST 29TH 1992

Now, a much more worrying proposition has raised its ugly head. As published in the *Banbury Guardian*, planning permission is being sought to create a rubbish tip in the sand quarry just above my beautiful Sanctuary. I shall object most strongly on the grounds of pollution to the underground springs from which all the water comes, the life source for more than 2,000 water birds and animals. The spring of pure water was my main reason for buying this land in 1989. It is now under threat from a RUBBISH TIP on land above me. This would poison the springs and be the end of a wonderful place for thousands of people,

Leaking acid batteries and asbestos

(mainly children), who visit here every year. Please God, don't let it happen! Paul Mobbs, an environmentalist, is launching the campaign to help save the Sanctuary. Bridget Barlow, a friend and well-wisher for years introduced me to Paul, who lives in Banbury. He got to work straightaway. *Friends of the Earth* have all the relevant legislation documents to hand. It will be a long haul, but please God give me the strength to fight this latest beastliness to mankind, children and animals!

AUGUST 31ST 1992

Paul printed a hundred letters of protest. All were signed by visiting families on Saturday and Sunday. Everyone is appalled at the application to tip waste above us. Paul made posters and letters within hours. He is a brilliant young man.

SEPTEMBER 3RD 1992

Paul came with photographs showing that, without doubt, tipping has been going on without planning permission. He questioned a young lady councillor from Oxford County Council, who came out to the Sanctuary at 3p.m., (she is in charge of the application for the Rubbish Tip). She said that they will seal the site off until planning has been agreed. To date there are now 2,000 signatures in all against the proposal. Well-wishers are queuing up to sign the petition against it.

Skip with lawnmower

More batteries & waiting skip

SEPTEMBER 7TH 1992

In a meeting with Mr. Grahame Handley at Cherwell District Council, he told me that the owner of the land had never been granted planning permission, only development permission. I still cannot see the difference in my view. It seems they have been contravening the law. Yesterday morning I went and stopped at the gate of the tip just as the owner drove up and introduced myself as the one opposing his illegal dumping above my premises. They are now asking permission to unload twelve skip lorries per day, full of builders' rubbish and tarmac road metal etc. at £52 per skip, adding up to £600 per day. It is big money and that is what all this is about — money, money, money, — with no regard whatsoever for the consequences to the environment nor for the Wigginton Water Reservoir, also beneath the tip. At 1.30 p.m. the Waste Disposal Officer for Oxfordshire County Council telephoned to say he was coming out to see me in order to look into it. It had been brought to his notice by another skip hire firm who thought he would be interested in the newspaper reports of September 4th. Indeed, he was interested, in a straight and honest manner. He can order the digging up of any alleged noxious materials above my beautiful Sanctuary. His visit lasted four hours, during which time he took a statement from Paul, of *Friends of the Earth*, as the person who took the photos of the acid/lead batteries lying in a heap on the ground at the quarry site on August 29th 1992. He saw and photographed six skips being unloaded. No-one can argue with this evidence.

SEPTEMBER 10TH 1992

The Waste Disposal Officer took me to the offending site so I could point out exactly where I saw them tipping and we did find evidence to support.

I took my protest package into M.P. Tony Baldry's office with the two thousand names and addresses so far on the petition.

Wendy has a new collie puppy to help her with her sheep; we are also training the pup not to bite or harm any of the many babies we have around at all times.

Thank you my duck!

Greenwatch

OXFORD MAIL — Tuesday, September 22, 1992

ANDREW STRANGE FOCUSES ON THE ENVIRONMENT

Fears over quarry pollution prompt an official probe

AN official investigation is underway after allegations of illegal dumping in a quarry which could threaten thousands of animals and birds at an Oxfordshire sanctuary.

Mabel Warner, owner of the Wigginton Waterfowl Sanctuary, claims that old oil drums, car batteries and lawn mowers have been left in the quarry which is being filled with soil to return it to farmland.

But the owners say they can't understand what the fuss is all about and have denied that anything apart from soil and clay has been deposited there.

Oxfordshire County Council experts have been digging at the three-acre site to see if anything is amiss and the National Rivers Authority is analysing water samples from a spring which runs to the sanctuary.

Two thousand people have signed a petition opposing an application to allow inert builders' rubble and other inert material to be dumped in the quarry.

The application by the owners will be considered by Oxfordshire County Council's environment committee tomorrow.

Children from local schools have written letters to the committee urging them to turn the plan down.

Mrs Warner looks after 2,500 birds and animals at the nearby sanctuary which includes 16 ponds. She fears that the permission will lead to more problems.

She said: "It will kill my animals and birds if they let this go ahead. It will destroy the sanctuary and spoil thousands and thousands of visitors' enjoyment of it."

Investigating allegations

Photographs allegedly showing items illegally dumped at the site have been taken by environmental investigator Paul Mobbs and sent to councillors.

Suzi Coyne, minerals planning assistant, said: "We are investigating the allegations that polluting substances have been deposited on site. If they are found to be substantiated we will take enforcement action."

Rob Sutton, of Brown & Mumford, acting on behalf of the owners, said nothing illegal had been dumped.

He said that the photographs taken showed batteries used in an old digging machine which have since been taken away and the oil was used for the same machine.

"We are a little mystified by this. We have been operating at the site for a little over a year with a licence to tip soil and clay.

"The county council monitor it...

RIGHT: Mabel Warner worried about wildlife sanctuary

BANBURY GUARDIAN Thursday, September 17, 1992

Threat to animal sanctuary

■ FLASHBACK: Our story of September 3, breaking the news of the threat to the sanctuary at Wigginton.

Don't ruin the wildlife

The following letter is a copy of one sent by the reader to Cherwell District Council:

I HEAR you are thinking of letting people put a rubbish dump next to Wigginton Heath, of all places.

If only you could cast your minds back to when you were young, you will remember the thrill of holding rabbits, feeding ponies, watching swans and stroking sheep.

If you put a rubbish dump next door to Wigginton Heath, you will ruin the wildlife that lives there.

Children everywhere come to experience the excitement of the creatures, think of the things you will be ruining.

Take a trip down memory lane, didn't you like seeing and feeding animals?

Children's dreams will be killed by one simple mistake. The mistake you made by allowing a rubbish dump to be put next to a wildlife sanctuary.

Simply, what I am trying to say is, Please, please, please don't allow a rubbish dump to be put next door to Wigginton Heath.

Aimee Stiff, Hillview Crescent, Banbury.

FRIDAY, SEPTEMBER 4, 1992

BANBURY Citizen

Somebunny loves them ...

● Amy Ryan, aged eight, sister Lucy, aged seven and eight-year-old Nicky Fox cuddle these fluffy bunnies at Wigginton Waterfowl Sanctuary.

The centre which attracts thousands of schoolchildren each year could soon close because owner Mabel Warner claims a proposed rubbish tip near the sanctuary will pollute her ponds.

For the whole story, pick up a copy of this week's Banbury Guardian.

■ SHOWING SUPPORT: Kelly Ward signs her name on the pe-

September 12th 1992

I received a letter from the Oxford councillor, who sent a copy of the application for tipping inert materials only. However, she failed to send a copy of the later application with the addition of builders' rubbish slates, tiles, metals etc. printed on it. This I had already seen in my interviews with Mr. Handley of Cherwell District Council.

September 13th 1992

Central Television cameras came to the Sanctuary to film the dump for inclusion in today's news bulletins.

September 14th 1992

I attended the Hook Norton Parish Council meeting at 7.30pm and presented my photos. Councillors will meet at the Sanctuary at 6pm on Tuesday to inspect the tip. They are concerned about their water supply, (the reservoir being beneath the tip). One hundred and fifty wonderful letters from the children of Croughton Middle School have been brought here by Betty Hanff. I photocopied them, sending them to Oxfordshire County and Cherwell District Councils, costing me a lot of money.

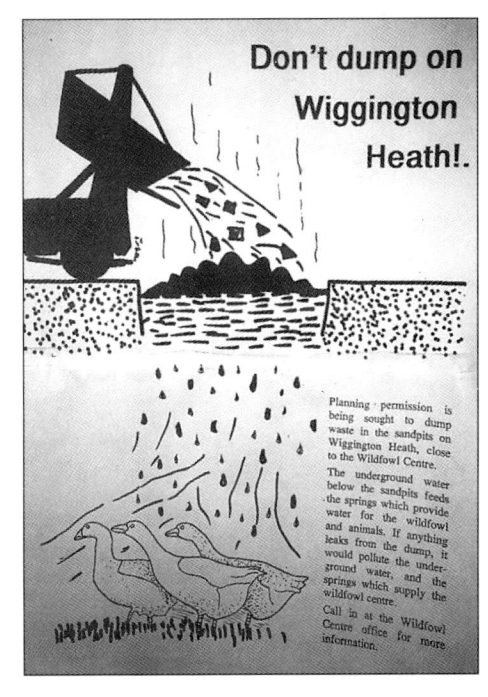

Paul Mobb's poster

September 15th 1992

I went to Cherwell District Council offices and saw Peter Williams who told me that only Mike Buxton and himself had seen the offending photographs and objections to the tipping of the rubbish. This was a week after my visit to Grahame Handley, who has gone on holiday until after the council meeting. I learned that their meeting was due to be taking place the day after Oxfordshire County Council was due to give its decision on the legality of the application to tip, (September 23rd).

Hook Norton councillors met here and went to the top tip with the owner's permission. I took them to the exact spot where I saw the lorry emptying (non-inert) waste and being covered up immediately by a JCB in front of my eyes, (three men with machines). When the owner and I met at his gate, he had invited me over to see the actions. I had neither the time nor the inclination as I had the urgent meeting with Grahame Handley in Banbury, which had taken as long as five days to arrange.

September 16th 1992

Arrived at Wigginton at 9am. Bridget was there bearing the rest of the petition from Hook Norton post office. I phoned Trevor Coles and was able to speak to John Weir, who assured me they were going to excavate up there at the tip.

Petition of 2000 signatures

SEPTEMBER 17TH 1992

In general, Wendy is a little unhappy about helping here, and has decided to spend more time away. We shall, of course, feel rather short-staffed and shall now urgently have to find some good homes for a proportion of the troublesome stock.

SEPTEMBER 18TH 1992

Trevor Coles and John Weir, the Oxford councillors, escorted a JCB to the site which then dug to investigate. I went up, but was told no-one was allowed, especially without protective footwear and hard hats. I visited again later staying on the roadside, but all I could see from that distance was soil coming out of the JCB bucket. They did not get over as far as I saw the tipping on Monday, September 7th in the presence of the owner himself. He told me that they were not tipping anything but soil and clay and I asked what it was they were doing with white plastic sheeting and evergreen being tipped and covered up immediately by his Drott shovel before our very eyes. I'm waiting patiently for results now. If nothing is found there, either whatever was there had been dug up again, or the JCB did not dig deep enough, or in the right spot. Lots more letters and pictures drawn by children have arrived. Another coach-load of little American children came this morning, but it rained most of the time so was not too pleasant for them.

SEPTEMBER 19TH 1992

Lots of support for the cause. My new neighbours at the Lodge came to say they had also written letters objecting to the tipping.

SEPTEMBER 20TH 1992

Stephanie is two today. Next Tuesday Micah will be eight. A nice few visitors today, in sympathy about tip. I gave away four of my older fluffy rabbits, as I worry about keeping them happy and their coats free from knots during the winter.

SEPTEMBER 21ST 1992

I visited Cherwell District Council offices to ask for the copy of recommendations and representations made to Oxford County Council. C.D.C. are voting against the tipping of anything on the site. Later I rang Trevor Coles, the Waste Disposal Officer. It seems it will be up to O.C.C. whether planning permission is refused. (Please God it is.)

"PLEASE don't let these creatures die!"

SEPTEMBER 23RD 1992

Today I went to County Hall, Oxford, for the meeting of the planning committee where the case was being discussed earlier than originally anticipated, due to its urgency. Norman Matthews spoke for me, also Paul, (*Friends of the Earth*), on the geology aspect. Although it was down for approval the whole committee voted against it, deferring their decision until November 4th pending a thorough investigation below ground. The Council are forcing the owners to dig up what they have been burying in the quarry, all at their own expense. Victoria Kingsman rang up from Horspath to hear the good news.

Swan Lake

I was invited to *The Mill*, Spiceball Park, by Royston Muldoon and Suzi Broughton to see a performance of *Swan Lake*. Last week they came to the Sanctuary to take photographs of swans and returned next day as they had decided to base their story on Swan Survival and problems with the quarry tip. They took a set of the photos to display in the foyer showing the story of Samantha and the threat to her and the rest of the birds and animals from the pollution. The performance was videoed by the husband of a teacher from Overthorpe School, Banbury.

Garden waste with slug pellets

SEPTEMBER 26TH 1992

Philip Brown, a neighbouring farmer, rang to say he had found the photos that had been taken on September 30th 1991 up at the sand pits. He had been worried even then as to what was being dumped there. This is just another small piece of evidence of illegal tipping, the shots showing a skip load of discarded plants and plastic pots etc.; also slug pellet containers scattered around, waiting to be covered up.

SEPTEMBER 27TH 1992

Quite a few visitors today, mostly supporters. I have managed to give away all of my Angora rabbits and a few big ducks to good homes. There's just Danny to go now, then, together, Rodney and I can manage to look after the remaining stock, having got rid of the difficult ones.

There is a substance, looking like oil to me, appearing on many areas of the centre of the grass area and also in the chicken runs.

Digger leaking oil

OCTOBER 12TH 1992

The National Rivers Authority came recently to take samples from my water springs and I am still awaiting results of their tests.

There's some horrid brown stuff oozing from the ground in the pony paddock today. I'm quite worried about this and really want to know what it is.

The owners of the quarry are currently tipping only soil/clay in the pit as they had previously been granted permission so to do. It would have saved a lot of trouble if they had stuck to the original conditions.

Teal ducks were swapped for call ducks and everything looks happy at the Sanctuary. The bank manager came, and was pleased with the way things have gone so far and is willing to help further in the future.

OCTOBER 18TH 1992

Betty Hanff of Croughton School brought a £1000 cheque raised by the school from sales. They are all helping to save the Sanctuary.

£1,000 US thanks for Mabel

AMERICAN children, going back to the States because of the impending closure of RAF Upper Heyford, said goodbye to a favourite place on Tuesday with a donation of £1,000.

Pupils from Croughton USA School presented their gift to Wigginton Waterfowl Sanctuary where owner Mabel Warner will refurbish some of her aviaries with the money, raised at the school's shop.

The children have been keen supporters of the animal refuge for several years and last summer wrote letters to the county council backing Mrs Warner's campaign to prevent adjoining land being turned into a rubbish tip.

Although the applicants, Smiths of Bloxham, were recently given approval to in-fill the site Mrs Warner says she will be keeping a 'close eye' to make sure the company stick to stringent conditions and does not cause pollution to nearby springs which feed her 16 natural ponds.

After presenting the money, pupils enjoyed a trip round the sanctuary which has several new residents including two Vietnamese Pot-Bellied pigs and a beautiful barn owl.

It will be just one of the places the pupils will miss when they have to fly back to America.

Heather Bedford, aged 13, has been living over here for the past four years but is returning to Georgia on June 12.

"I'm going to miss the countryside the most," she said.

Stephen Guy, aged 14, will be leaving his home country – he was born in Ipswich and apart from living in Germany for a few years has lived in England all his life. He will travel back to the States with his American father and British mum next year.

"I think of England as my home - I'm really going to miss living here," he said.

But 14-year-old Raschelle Corsey, who has lived here for two years, says she is looking forward to going back.

"There's plenty more to do there," she said.

■ ANIMAL AID: Teacher Betty Hamaff and pupils from Croughton USA School present sanctuary owner Mabel Warner (right) with their £1,000 farewell gift.

A pair of Saxony ducks came to live with us, quite old, but still sprightly. I sold two miniature ducks, a birthday present for a husband from his wife.

I received some letters back from councillors. A fifty pound cheque came from *Friends of the Earth* to help fight the tip threat, also a letter from John Weir, not very convincing. He says they have investigated three quarters of the tip, but I still believe they have not dug deep enough behind the mound of earth that was created.

October 22nd 1992

Mr Lowther, from the Environmental Department of C.D.C., came to investigate the noxious substance that had appeared in the pony paddock.

A letter from Grahame Handley — encouraging.

October 27th 1992

I sold a "mini bunnie" to Mel, who named her Mary Poppins; she was disappointed with her first bunny, which grew up and had a few behavioural problems. It quite blotted my copy-book.

November 4th 1992

Rodney is building four more aviaries against the back fence behind the mobile home, a fairly sheltered corner. It will house pheasants, blue magpies and other big birds, flying above, with bantams and rabbits living below at ground level. We are using off-cuts and all second-hand materials, so as to keep the cost down to a minimum. It is an added interest for visitors.

November 5th 1992

A group of five-year-olds from Overthorpe came to visit and loved getting nice and muddy.

Alison, their teacher, brought them but was dreading what a few mothers would say about their clothes. Yet the children looked so happy with their glowing cheeks and bright eyes. Before leaving Alison popped back to say I should write to Persil to say what a perfect advertisement it would make on television!

NOVEMBER 10TH 1992

I sat for about an hour in the summer house up at the top of the Sanctuary just looking and thanking God for everything here. *"How beautiful it all is!"*, I thought, as I silently thanked Dennis, Rodney and all who make it possible. The sun was shining on this lovely autumnal day, the oak trees were golden bronze, leaves falling gently to the ground making a carpet of colour. Four little Dutch bantams were happily scratching around for grubs. I watched this mother and her three little girls who decided to live among the ducks and geese and who perch on top of a particular duck house every night, instead of going inside for protection from the wind and weather. Two Phoenix fowl cockerels perch in the oak tree at night, high enough to escape the fox. I saw many wild birds today: hedge-sparrows, blackbirds, the funny little jenny wren, chaffinches, wagtails, magpies, crows, ring-neck doves — lots of wild birds come to feed around the pens.

Some white doves were brought by a man who said they would have to be destroyed if I couldn't take them in. Oh, they are so lovely, — pure white. We put them in the big flight pen to be released later when they have settled down and know they are welcome here. The many pigeons and doves who live here with the chickens, sharing their houses, are all very happy together.

I sold six Marans, (and one cockerel), for £15 each, point of lay, my specials. Hopefully, they will reproduce a high proportion of females.

NOVEMBER 11TH 1992

White doves settling down well.

NOVEMBER 12TH 1992

Five pound cheque from *Friends of the Earth* — many thanks.

NOVEMBER 15TH 1992

A mother duck came waddling out of a hedge with six baby ducklings, – quite unexpected so late in the season. Three died despite putting them under an infra-red lamp. Luckily, she was spotted, as they would all have perished from predators.

NOVEMBER 22ND 1992

A school group came from Coventry and was so successful the headmistress promised to publicise us around her school area. One little girl slipped and hurt her back, but quickly recovered; we must fasten more coverings of wire netting on the bridges to prevent any further mishaps.

A visitor brought me a box of chocolates as a present for me for taking in baby bunnies. A kind lady from Shenington brought in a wandering lost duck and gave £25 to help feed the animals.

Later in the evening Rodney carried out the work on the bridges.

NOVEMBER 28TH 1992

Very sad news today — my mother, Mary Taylor, died aged 91 years. She lived a wonderful life and was a good example to her whole family and everyone she met. Her death was very sudden and yet very peaceful. The greatest loss is going to be for my sister Peggy who had cared for her so devotedly for eight years, since going to live at Murcott House. The funeral will be at Charlton-on-Otmoor Church on Friday, December 4th.

Hooky the Hedgehog

DECEMBER 5TH 1992

Last weekend someone brought a little hedgehog. I named him Hooky, the same as the one I had several years ago, after picking him up off the road at Hook Norton. At the time I was in the flower trade, and the earlier little Hooky licked all the jam from inside a large jam bucket that we had just collected for use with the flowers. From then on I fed him bread, jam and milk when he came every morning to the back door of our cottage at Brailes.

Today a little boy came from London, together with his mother and some other children. He nursed little Hooky handling him wrapped in one of Granny Taylor's many knitted blanket squares. His mother said she thought this place was absolutely wonderful and her children would never forget the experience. The other children handled tame mice and baby rabbits.

DECEMBER 9TH 1992

Paul came to take his poultry back to his school, at Neithrop in Banbury, as their new building is now ready. Lucky, lucky Neithrop!

DECEMBER 16TH 1992

Early this morning the little hedgehog had disappeared from his cage. Later on I heard a little baby rabbit squeal and dashed over to see why. Lo and behold, Hooky the hedgehog was buried in their warm nest and the poor little things were getting pricked. He evidently lifted up the wire grille at the front of the cage and so must be quite strong as there is no other way in. Richard, from Milcombe, brought a mouse for release. My friend Bridget Barlow, from Hook Norton, called in with a large bag of pet rabbit mix as she had just lost her pet rabbit, seven years old.

DECEMBER 25TH 1992

Nice few cards for the animals. Alex Potter and Cliff, her dad, brought photos of the three friends handing over a cheque £45 for our friends, the animals. Bless them! *See colour plate.*

John Ekers, with his three little girls Naomi, Hannah and Sadie brought a tin of biscuits from Jackie, his wife, — for us, not the animals, this time!

DECEMBER 28TH 1992

Another young barn owl brought in, aged about three months. She had been found in a garage and raised by a Middleton Cheney couple. They had had the parent owls for twelve years and thought they were past breeding. The unexpected arrival of this baby owl presented them with problems as they had no room for another flight pen. I accepted her, complete with registration ring number.

JANUARY 1993

Rod, with a little help from me, finished off the aviaries. There are just the perches to fix now.

JANUARY 28TH 1993

Samantha the Swan died; she just went to sleep peacefully on the pond, very gracefully; she died as she had lived, with dignity and she must have been a great age. I hope it wasn't pollution.

We finished one lot of aviaries, naming them the *Croughton Corner* as the £300 raised by Croughton School went towards buying the building materials.

FEBRUARY 15TH 1993

I purchased three pairs of ornamental pheasants from Mrs. Sue Easton of Angus, Scotland, who breeds many varieties — one pair of Monal Himalayan, one pair blue-eared, and one pair silver. *See colour plate.*

MARCH 13TH 1993

I went to a farm sale at Hailey, and bought a corn trailer, wire panels, alkathene piping for water and for the adventure playground. Today Rod held the fort and their were lots of visitors.

MARCH 14TH 1993

Sunday and a lovely spring day. Lots more visitors, many returning again, like the two children who wrote in visitor's book *"the best place in all the world."*

Olwyn, the barn owl, has been adopted by little Catherine Pearce from Bloxham. Barn owls seem to be a great favourite with many visitors. It is so nice for them to get really close up to Olwyn; she is very calm now, getting used to the many people who look in wonder at her, and return her deep, deep searching gaze.

Our dog Annie, a seven-year-old Golden Labrador, who had been suffering from a terminal disease, finally had to be put to sleep. It is very sad and we miss her terribly; we received a nice letter of sympathy from our vet, Kate.

One of my new peacocks has decided to go and live on Marcus's barns next door. Twice I endeavoured to persuade him to come home and straightway he flew back again. I hope he comes home when he gets hungry. They are such magnificent birds and I hate to have them caged.

MARCH 18TH 1993

I attended a seminar at Cotswold Wild Life Park to learn how to promote my attraction to the schools through the Southern Tourist Board. I learnt a great deal: it seems I have the kind of place schools will enjoy as much in the future as they do at present, after they have been made fully aware of our presence in the area. Phil, the speaker, is the same gentleman who organised the Welsh Flower Festival in 1991. There is no doubt he knows what teachers need and want, having been a teacher for twelve years. He is now an advisor, lecturer and planner, a man of many talents.

I was invited to the *Crown and Cushion,* Chipping Norton, where a social evening was held to meet the Southern Tourist Board management and staff. I am now joining them, following the collapse of the Thames and Chiltern Tourist Board. I was pleased to see Stan Bowes, who is still the representative for this area. It was a pleasant evening all round.

Once more, lots of spring chicks hatching now: Cochins, Blue, Buff, & Partridge; Pekins, Black and Blue; Marans; Rhode Island Reds; Wellsummers; Phoenix; O.E.G. and Silver-laced Wyandottes; Polands; just a few of each. I bought in young day-old ducks from Colin and Tina Chester of Greatworth.

APRIL 4TH 1993

Two Sebastopol goslings hatched under mother goose.

APRIL 8TH 1993

Zoë, the Soay sheep, gave birth to one male lamb. Many more chicks.

APRIL 9TH 1993

Good Friday, — rain all day — a good attendance of visitors despite the weather. We are four years old today!

APRIL 12TH 1993

David Shepherd, our vet from the West Bar veterinary hospital, borrowed Easter chicks for his church. All the children love the chicklets. The most popular place is the cuddling area.

APRIL 17TH 1993

A Tortoise egg brought in so we can try to hatch it in the incubator, if we are lucky.

The blue magpie has laid seven eggs, so I promptly removed the cock bird to another aviary in order to avoid last year's disaster.

April 18th 1993

Teachers from Harnham School made a two-hour journey from Salisbury, Wiltshire, to collect an incubator and twenty-four rare breed eggs as previously arranged. They are planning to bring approximately seventy children to visit when the chicks have hatched.

April 19th 1993

Two little female Vietnamese Pot-bellied pigs have been brought free to the Sanctuary. Their owner, Jackie, had too many and couldn't sell them, so I took some off her hands. They arrived complete with an animal movement licence and are quite adorable. I can see now why people generally go into raptures about them. They are approximately seven months old, about twelve inches high by eighteen inches long, jet black with little snub noses and sharp black eyes — very friendly. I have named them Dotty and Dora.

April 20th 1993

A small boy brought me an Easter egg just because he loves my place so much — lovely.

Victory!

April 23rd 1993

I attended the Oxfordshire County Council meeting to decide the future success or decline of my Sanctuary. Oxfordshire Caring Countryside Council has been discussing the approval of using the Quarry above the Sanctuary for a waste dump. Planning permission was given to fill with *soil and clay only*. Victory!

Now I am having all of my ponds cleaned out. Slurry from the bottom is being taken up to the top field out of sight, the clay dug from the bases piled and compacted around the outlet sides of each one. Hopefully, the shallower ponds will not now dry up in a drought summer as they once did. The cost for digging out is proving quite expensive.

June 13th 1993

One pair of barn owls and one young one each ringed with I.D. numbers arrived via Marcus Ridley from Somerset. Eventually they will be released from cages to stay around the area.

Dedicated work needs supporting

DURING a brief visit to Oxfordshire, I by chance spotted a road sign to the Waterfowl Sanctuary and spent a most interesting couple of hours there.

I could not help but think how fortunate the people in the Banbury area, and indeed further afield too, are to have someone devoted to the welfare of not only waterfowl, but also other species of wildlife, amongst which I noted two owls.

It became obvious to me that the dedicated work of Mabel Warner and her helpers could be further developed and enhanced with more local support.

Perhaps your newspaper could encourage visits, including organised parties and especially parents and their children for a never to be forgotten treat.

I.M. Lishman, Sandygate, Isle of Man.

Letter in Banbury Guardian

"It's like baby Jesus, isn't it?"

JUNE 17TH 1993

Hook Norton Primary School came today. Sitting on the row of little chairs, one small boy, after being handed a tiny yellow duckling swathed in a blanket looked up at me and said: *"it's like baby Jesus, isn't it?"* What a lovely thing to say! Ducklings are wrapped in blankets because they tend to leap around, whereas chicks and bunnies are relaxed on children's laps.

JUNE 22ND 1993

Two little girls from a school at Overthorpe held a sale and made £38 for the Sanctuary funds, selling lavender bags, sweets etc. that they had made.

JUNE 23RD 1993

A great-crested grebe was brought in, entangled in a fishing line. I picked him up and straight-away he pecked me in the eye, just missing my eye-ball.

JUNE 24TH 1993

The grebe made a good recovery and has been returned to the Brackley lake today.

The Rat in a Hat

JULY 8TH 1993

My most urgent problem at the moment is that of rats. A pest control surveyor, when he first came, promised me he would put the bait deep into the rat runs and block up the entrances securely. Unfortunately, when the officer came to set up the bait locations for the extermination of the rats, he failed to cover them up properly as promised. Then the hens promptly scratched the pellets out on to the surface of my rare breed poultry pens, containing very valuable birds, the poison needed to be removed as quickly as possible, grain by grain. I was extremely concerned and angry about the situation. Pest control was immediately alerted and two officers came at once to pick them all up on their hands and knees. Baited sealed boxes had been placed inconspicuously around the farm. As a result fifteen rats were killed.

A few days later the unexpected happened! A lady visitor picked up a near-dead rat and put it into her sun hat, brought it to me and said: *"I found this poor little animal, it looks ill"*!! I immediately told her to go quickly and wash her hands and advised her to burn her hat. She explained that her hat had been bought abroad and was very special. Her husband, also very concerned, wondered whether there was a fumigant they could use so that after treatment she could wear the hat again? I gave them the telephone number of the pest control firm so they could contact them for advice. My best thanks for their tolerance in an unusual situation – I wonder if I'll ever know if they saved the hat.

A peacock, in full plumage, died from heart failure. Carl Hutt, a taxidermist, quoted a special price of £150 to preserve it for me.

JULY 1993

One dear lady visitor brought two girls but even though it was raining she would not enter one barn where we keep tame mice. The children loved holding them, but she was petrified. I explained that these pet ones do not dash around like wild ones, they are Australian fruit-bat mice with suction feet and they cling to fingers! They are totally harmless. Another mother brought two small children. She explained that she could not comfortably stay inside with animals but came so the children could enjoy cuddling the many babies. On this occasion I made a special exception and supervised the four little children myself as there were few visitors on this rainy day.

I foolishly offered to mate a bad-tempered doe rabbit with one of my bucks. She was not responsive immediately so I suggested she leave her overnight. When I looked at them in the morning, the beastly female rabbit had broken the front foot of my male – bitten right through it. The vet reports a compound fracture and part of the foot will have to be amputated. The treatment cost £42.42, of which the owner reluctantly contributed £20. My poor buck, meanwhile, was out of action, bandaged up and feeling very sorry for himself.

A group of strong young men offered to do voluntary work in their own time at the weekends. They arrived on Saturday August 7th. and by Sunday evening had built most of the owlery with timber and offcuts. Their wives and children all came to lend helping, willing hands. The next weekend the wire was fixed and the job completed. Warrant Officer Ewer, who was in charge of the group, has since moved to Cornwall. I believe that the project formed part of a goodwill community gesture to compensate for the noisy aircraft currently flying from their airbase.

July 15th 1993

A nest of baby rabbits was born to a very small mother, the tiniest so far.

July 22nd 1993

Finmere School visited and also North Hinksey School. A heavy drinking bowl was kindly donated for the animals. Charlbury Primary schoolchildren raised £120 from a sale they held in aid of the Sanctuary. I must try to offer the schools more in the future.

July 24th 1993

Two packages of thank-you letters arrived from children of Emmanuel Christian School, Oxford also lovely drawings from Hook Norton School.

August 4th 1993

I used the money given to the Sanctuary by Croughton USA School to buy concrete blocks to line part of the feed shed making it mouse and bird proof. Maurice is doing the work, calling it the Croughton Cabin for the display of letters and paperwork from the many schoolchildren.

August 30th 1993

Six weeks later the three tiniest rabbits still stay very small, about the same size as my three female Mabel's Midgets and I hope to breed from them in the spring.

August 31st 1993

A fantastic number of visitors have come throughout the school summer holidays this year.

Maurice finished my Croughton Cabin schoolroom.

Dennis harvested wheat and barley, enough to feed the stock through until 1994. Valuable help from Rodney, Wendy and Martin Cherry.

September 5th 1993

My elder daughter Jackie is 37 today.

September 11th 1993

Den and I thoroughly enjoyed a short holiday in Scotland, travelling with Jeff's Coaches. It was lovely: to the Western Isles, and the Isle of Skye.

September 18th 1993

Returned home, everything fine; Rodney and Wendy have managed very well.

Vietnamese Pot-bellied pig gave birth to a female offspring: quite unusual to have just one.

"THE COTSWOLD STANDARD" PHOTO FEATURE
September 20th 1994

You don't have to be mad to work here – but it helps!

Aylesbury duck & black pigeon

Australian Fruit Bat Mice

MABEL WARNER readily admits she is something of an eccentric. She has to be, she explains. It helps her to pursue her passion with such enthusiasm.

How else would you consider the activities of the 63 year-old farmer's wife who is prepared to work up to 16 hours a day, 364 days of the year? "But I enjoy every minute of it", she will tell you.

And she will give you 3,000 good reasons why her work will continue. For that is the number of birds and animals living in peace on and around the 16 ponds on her 22 acres just outside Hook Norton.

The idea started when Mrs Warner realised no-one was prepared to look after unwanted ducks and geese. The ex-florist decided to establish her own sanctuary. By the time she launched the idea in March, 1989, she was running a rescue centre as well.

Today the sanctuary at Wigginton Heath attracts up to 17,000 visitors a year and Mrs Warner "just keeps my head above water."

But Mrs Warner has made an outstanding contribution to the conservation of rare breeds and wildlife, and the site is a haven for a great variety of birds and other interesting and unusual creatures.

The original idea of setting up a sanctuary for ducks and geese has expanded to such an extent that today visitors – especially parties of schoolchildren – are delighted and amazed to find Highland cattle, sheep, pigs, peacocks, mice, owls, rabbits, goats, a Shetland pony, guinea pigs, etc.

"And I know that many of them would be dead but for my work", said Mrs Warner. "It is that knowledge which has kept me going".

She has such a reputation for her rescue work that unwanted pets arrive from an area taking in Wales, London, the north of England.

Often she goes out and about collecting birds and animals she has heard will be put down. Like the peacocks she saved from being shot when their owner moved across country to a smaller home. The six goldfish (since risen to 150) she saved from a pond which was due to be bulldozed. And the fully grown swan which she found when a cygnet with thorns embedded in its chest.

Mrs Warner appears to do her best to dispel the theory that one person cannot be in two places at once as she carries out a non-stop programme of activity as soon as she arrives at the sanctuary from her home at nearby Enstone.

And if, as she claims, she is eccentric, then some of her eccentricity had rubbed off on her guests. She has a ring-necked parakeet and an Angora rabbit who have been carrying on a love affair for four years. The sight of them sharing a carrot, one at either end gradually munching towards each other, will never be forgotten!

"Yes, of course it's hard work, but it is also fun", said Mrs Warner. "But when you look at the faces of the children who come here, hear their expressions of joy, read their letters of appreciation, I know it is all worthwhile."

○ The sanctuary is open seven days a week, round the year (except Christmas Day) from 10.30am to dusk, or earlier by appointment. It is just outside Hook Norton. Just follow the duck signs from the A361 or the B4035.

Black Indian Runner Ducks

Australian Black Swans

A mallard drake was brought to me after undergoing surgery. He was found by Margaret Bullock near Banbury. The vet inserted a pin a into a severely crushed leg. It became gangrenous and had to be amputated. We have kept him penned up for three months and he is just about ready to be released back on to the pond with other ducks. He gets around very well with one leg only.

West Bar Vet. Tony Roberts, Margaret Bullock and myself

OCTOBER 23RD 1993

I sold five of my Mabel's Midget rabbits: two went to Oxford and one each to Northampton, Milton Keynes and Evesham. These midget rabbits are unique to the Waterfowl Sanctuary. They were bred very selectively in the early days from an unusually good-natured Netherland Dwarf buck and the dillon of a litter of small cross-breeds. I later introduced an extremely placid Poland into my stock. All of my Midgets are miniature strains of several recognisable larger breeds.

Australian fruit bat mice are being trained by my own hands to cling on to my fingers. From the day they are born I pick them up and the children love them. I have thirty babies at the moment, brought originally from Jonathan at Hook Norton. *(See colour plate.)* Rod has just finished a second block of new aviaries and concreted floors with my help. Melody and Micah came after school and they loved sploshing around in wet cement in wellie boots – not a good thing to do, really!

OCTOBER 24TH 1993

I put young stock in one of the new cages with Michael's help, and stocked second pen with laying hens — (Michael and Wendy make a good pair.)

NOVEMBER 1993

A good small article in the *Banbury Cake*, featuring Birthday Girl, Eve, with some of the Fruit Bat Mice.

Eve's as quiet as a mouse

CUDDLING these little cuties kept birthday girl Eve Crackle as quiet as a mouse at Wigginton Waterfowl Sanctuary and Rescue Centre.

Eve, of Bourne Lane, Hook Norton, enjoyed her seventh birthday treat taking a close look at the fruit bat mice born at the sanctuary six weeks ago.

They have been hand-reared by centre owner Mabel Warner, who is now looking for good homes for them.

Mabel said: "Altogether we had 29 baby mice, born to three mothers and two fathers, which I am keeping in an aquarium tank.

"They are absolutely fascinating to tiny, little children, who like picking them up and cuddling them."

Anyone who would like one of the baby mice, which are being sold for £2.50 each, can contact Mabel on 0608 730252.

A new-covered picnic area, which will be used for school parties, has just been completed at the centre, thanks to a £1,000 gift from US pupils at Croughton School before returning to the States.

JANUARY 26TH 1994

Permission!

Permanent planning permission has been granted for the Sanctuary as a tourist attraction, so the future is looking good!

Only a few trusted close friends know that as well as the Queen's *Annus Horribilis*, I also had a similar year, during which I faced the prospect of losing my beloved Waterfowl Sanctuary. A writ was issued against me for the sum of £32,000 for the rebuilding of the front wall of a house I sold in 1985 to a greedy, unkind foreign woman. In fact, her architect had written that the repair of an existing crack would cost no more than £1,000. I feel that two of the solicitors involved could have stopped the case right at the beginning from the evidence written in letters I gave into their care. Both chose to keep quiet until much money had been clocked up against me. I was granted legal aid. The case was closed after she accepted the original £1,000 that I offered to remedy the crack, but now it will take another two years to pay back the legal aid — causing me many years of unnecessary worry.

Additionally, a tip of rubbish was started right above my beloved Sanctuary, whence come the springs to feed the sixteen ponds. Being the eccentric I am, once again, I fought like a tiger to stop this and thanks to *Friends of the Earth*, the noxious dumping was stopped by the Council only an infill of earth and clay being allowed. This is quite acceptable to me.

I replaced my pair of beautiful black swans from a breeder in Norfolk at half the price I paid for my original ones. I live and learn. We had to shut them into a house one evening when temperatures dropped to below freezing with stinging, gale force east winds. I feared for their lives. In the evening Wendy and Rod helped me to get them in to their house, and the male (cob) raised his body threateningly to its full height. He was eye to eye with Wendy, nearly six foot tall, but she stood her ground and we captured him for his own safety. Normally the swans sail gracefully along together on their pond looking magnificently serene and extremely dignified with their bright red beaks, jet black glistening feathers and white tipped wings just protruding from their body sides. (The female is called a pen.) They originate from Southern Australia. (Imports are now banned, I believe). Within three months they become tame and will ask quietly for food, grain and attention.

February 1994

Schools are starting to re-book for 1994; four have so far taken up the incubator offer and are hatching eggs in class. Michael Page has some incubators at his home in South Newington and is overhauling them.

"The Magical Stone"

February 22nd 1994

Linda and Christopher, aged ten and seven, who live in Hillview Crescent, Banbury and their friend, Sarah Babbs, of Trimsbury Court, Oakwood, Derby, wrote and performed a play called *"The Magical Stone"* to friends and relatives and raised £8.53p for the birds and animals. I wrote a thank you letter from the animals. Bless the many kind children who help raise money.

March 10th 1994

Telephone call at home 10.30 p.m. — the birth of calf imminent.

March 13th 1994

Mother's Day. Calf born to Wendy's Limousin black cow; she is over the moon! She raised her from a 10-day-old calf together with another one.

First of the call ducklings hatched; thirty chicks from the incubators.

April 27th 1994

Four grey Micro Midgets born to Poppy, a good mother. Penny, her sister, has not been mated yet.

April 29th 1994

A mallard call duck hatched seven young on the first pond. Each day there was one less until all had gone to sad ends. I suspect, crows, magpies and jackdaws, always on the look out for vulnerable babies, are responsible. Nature can be very cruel.

May 14th 1994

Horspath School, one of the best supporting schools in Oxfordshire, came on yet another repeat visit. They have been several times with Lorraine Kingsman, one of their senior teachers, who is to become Headmistress from September this year.

May 16th 1994

Another micro midget mother gave birth to five babies: three white and two dark grey.

May 27th 1994

Two teachers came with the results of their classroom projects: Mrs. Gaffrey of Great Tew School brought along a successful hatch, and also, Sara Wade of Cutteslowe School, came with eleven chicks. *See letters at back of the book.*

May 28th 1994

Another good friend, Mrs. Ann Morton, headteacher of Finmere School brought back the incubator and chicks. Also, Mr. John Morgan of Bishop's Itchington School brought back eleven chicks, — once more very successful class projects. The schools really are enjoying the use of the incubators. They were supplied with eggs of the rare breeds of poultry. I'm finding low fertility in the early season eggs, but as the weather warms up better hatchings emerge.

Recently I have hatched bantams, hens and ducks, geese, five Brecon goslings, three grey Chinese, two white Chinese and two Sebastopol.

August 7th 1994

Following planning permission, a large, second-hand barn is being erected for young animals in the children's farm. Martin Cherry and Rodney are doing most of the work. They have taken it down from Enstone, re-designed it and put it up at Wigginton. The 60ft x 60ft barn has concrete floors, damp-proofing and part block walls, built by my brother-in-law, Maurice North, Mark and Martin. Winter conditions for the animals, thanks to them, will be much more comfortable.

Many more people are finding the Sanctuary and returning again and again. Well-wishers who come to bring fruit, vegetables and bread are especially welcome. They include Margaret and John Cheal from Easington Road, Banbury. Deborah, an American who lived in Deddington, sent a £25 cheque. Great helpers also are John and Margaret Radway, who donated some cockatiels, from which the first babies are now being hatched; also Enid, Annie, Dion, and friends from Wolverhampton.

AUGUST 1994

I have become a small distributor for hybrid point of lay pullets. They are faster growing than many of the rare breeds I keep and they sell for a much lower price. My preference is for Black Rock and Warren Shavers; also, speckledly-brown egg-layers. They are friendly and make good pets, producing many eggs, for folk who want to keep a few hens in their back gardens.

A visit from Lorraine Kingsman, from Horspath Primary School, her daughter, Vicky, and family friends. They are great well-wishers of the Sanctuary. A little note was gently put into my hand by Elizabeth Blake.

I have many letters from various schools who have borrowed my incubators, hatched eggs in their classrooms, then visited the Sanctuary.

"Waterfowl Sanctuary and Children's Farm"

As a result of a visit by Judy Warren of Slough, I have now definitely decided to add *"Children's Farm"* to our name. She is a freelance designer of fact sheets for schools and formerly a primary school teacher and is now designing reception class sheets for me. All printing will carry the new heading. Printer Richard Payne and his wife, Mary, are very patient with me, ensuring that everything is word perfect.

AUGUST 26TH 1994

We now have a wonderful crop of wheat in the barn, grown at Church Enstone by Dennis, my husband, combined by John and Martin Cherry, hauled by Wendy, May and Rodney, augured from the trailers by Den and myself. Everyone worked hard until past midnight to get the corn in dry, as the weather is very unpredictable. Straw is being baled and stacked, some on trailers waiting to be stored on the concrete floor in the barn at Wigginton. We should go through this winter with not quite so much worry as in previous years when we struggled to keep tarpaulin sheets on the bales.

Jimmy, the terrible Cockatoo

A lady and gentleman from Silverstone, Mr. and Mrs. Smith, brought me a Yellow Lesser-crested Cockatoo. He says *" hello Jimmy, hello darling"* as well as emitting a deafening screech, stopping all conversation. I put his cage, too, near the incubator and he chewed through the cable and cut off the power. Luckily for him we had fitted a very sensitive automatic cut-out. (None of the eggs has suffered as they hatched out later.) He loves crisps and when he sees anyone eating them he chatters and screeches. He came here because, when his owner became a grandmother, the cockatoo became possessively jealous of her cuddling the baby. Jimmy's agitation and noisy screeching became so bad she did not dare leave them in the same room together. Even so, his owner had a heavy heart when leaving him in my care. It is possible the cockatoo will return to her in future as he is tolerant of older children. Meanwhile, he is a great attraction, mainly to adults, inviting tickles by spreading his beautiful white and yellow wings.

Among the many wonderful well-wishers who frequently visit are Ted and Mary Jolliffe of Clapham, Bedfordshire who bring fruit and vegetables for the birds. They have a holiday caravan based just outside Banbury.

Also Nancy Bowstead of Wellesbourne distributes leaflets and also collects heavy-based feed bowls for the birds and animals.

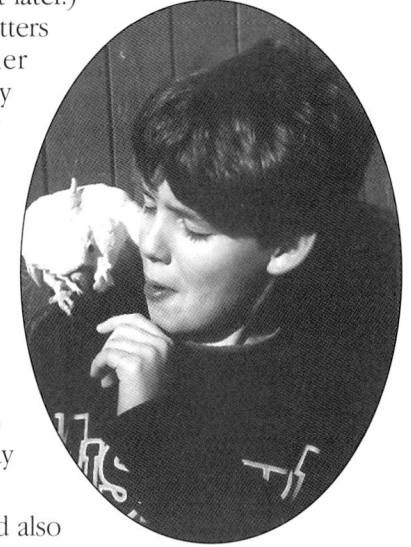

Micah with Jimmy

SEPTEMBER 1994

My bank manager has agreed to lend me more money for improvements.

SEPTEMBER 27TH 1994

A black tufted duck has been brought in: although it was shot through its wing, it is healthy and can swim. Antibiotics were given.

OCTOBER 15TH 1994

Maurice, Mark and Martin North are putting in the concrete floor of the barn. This is the one that some cowboys started to put up for me at Enstone four years ago, and it all collapsed. Now it has been re-designed by Martin Cherry and is now standing square and firm with good foundations. The fact that Martin lives nearby helps enormously.

NOVEMBER 5TH 1994

The Ponds are all dug out once again — it took three days. Barn floor complete, building progressing well. Good drainage all round.

DECEMBER 1994

Sibford Friends put on a concert and raised £31 for us. I wrote a thank you letter through *Banbury Guardian* as they did not sign a name or address.

DECEMBER 7TH 1994

The barn is finished all but for the large doors. Martin and Rodney are exchanging a few workdays. They sometimes use a kind of bartering system in working for each other.

JANUARY 19TH 1995

We had our one minute of glory when *"Rooftop Invaders"*, an entertaining wildlife film about jackdaws was shown on ITV tonight. Tim Shepherd of Chipping Norton filmed it, a *"Survival Special"*, for *Anglia Television* four years ago. I marvel at his incredible patience as one minute's screen time showing Rodney with Jacko, our jackdaw, took Tim a whole week to film! They gave us a credit in the unrolling scroll at the end.

FEBRUARY 1995

Rain, rain, rain – *"February, fill the ditch/ dike "*

MARCH 11TH 1995

Once again, chicks hatching: Cochins, Pekins, Silkies, Barneveldas, and Phoenix.

A large black lop-eared rabbit was brought in. So many are purchased at pet shops or garden centres as "dwarf lops" (meaning the ears, not the body, as so many people are led to believe). Nearly all baby rabbits sold from Wigginton are seen with their mother before they are purchased.

A couple who live at Bloxham visited, and were thrilled to bits with it all – amazed they hadn't been before!

MARCH 12TH –14TH 1995

Sunday, a lovely sunny day, lots of happy visitors all day. Monday, rain; Tuesday rain.

MARCH 15TH 1995

I purchased hatching eggs from Shropshire £10 per doz: Barneveldas, and Wellsummers.

MARCH 16TH 1995

Richard Valdambrini *(see colour plate)* finished the promotional video which he made for me free of charge, borrowing his equipment from a friend in London. They will be sent out to local schools free. A teacher from the Fairway Playgroup took one.

MARCH 17TH 1995

Gale force winds today – they keep visitors away.

MARCH 18TH 1995

The Fairway Playgroup teacher rang to tell me she is very pleased with video and will be visiting with the children shortly.

MARCH 19TH 1995!

Lots of afternoon visitors despite the snow, hail and rain. My nephew, Brian North, and Sue, his sweet wife, came with their Sunday school children and all enjoyed their visit. My policy is not to charge entrance fees to my many relatives and very close friends, but I found a fiver in the Thank You Pot after they had left. Thanks, Sue!

MARCH 20TH 1995

I attended a leaflet Swop Shop at *Hopcroft Holt Hotel* organised by Stephen Higham and Vickie Hope-Walker of Cherwell District Council; the attendance was disappointing. I was quite worried to see there was a stand advertising Grasshoppers Children's Farm, recently opened near Bicester. *(This closed down after a few years.)*

A bird in the hand

MAGGIE the blue magpie is one of the rarest pets around. The nine-year-old bird, whose natural habitat is in south east Asia, lives at the Wigginton Heath Waterfowl Sanctuary, owned by Mabel Warner (pictured). See the sanctuary in TV's Roof Top Invaders next Thursday (January 19) at 7.30pm on Central.

Thank you my duck!

A sample of the reception class fact-sheets

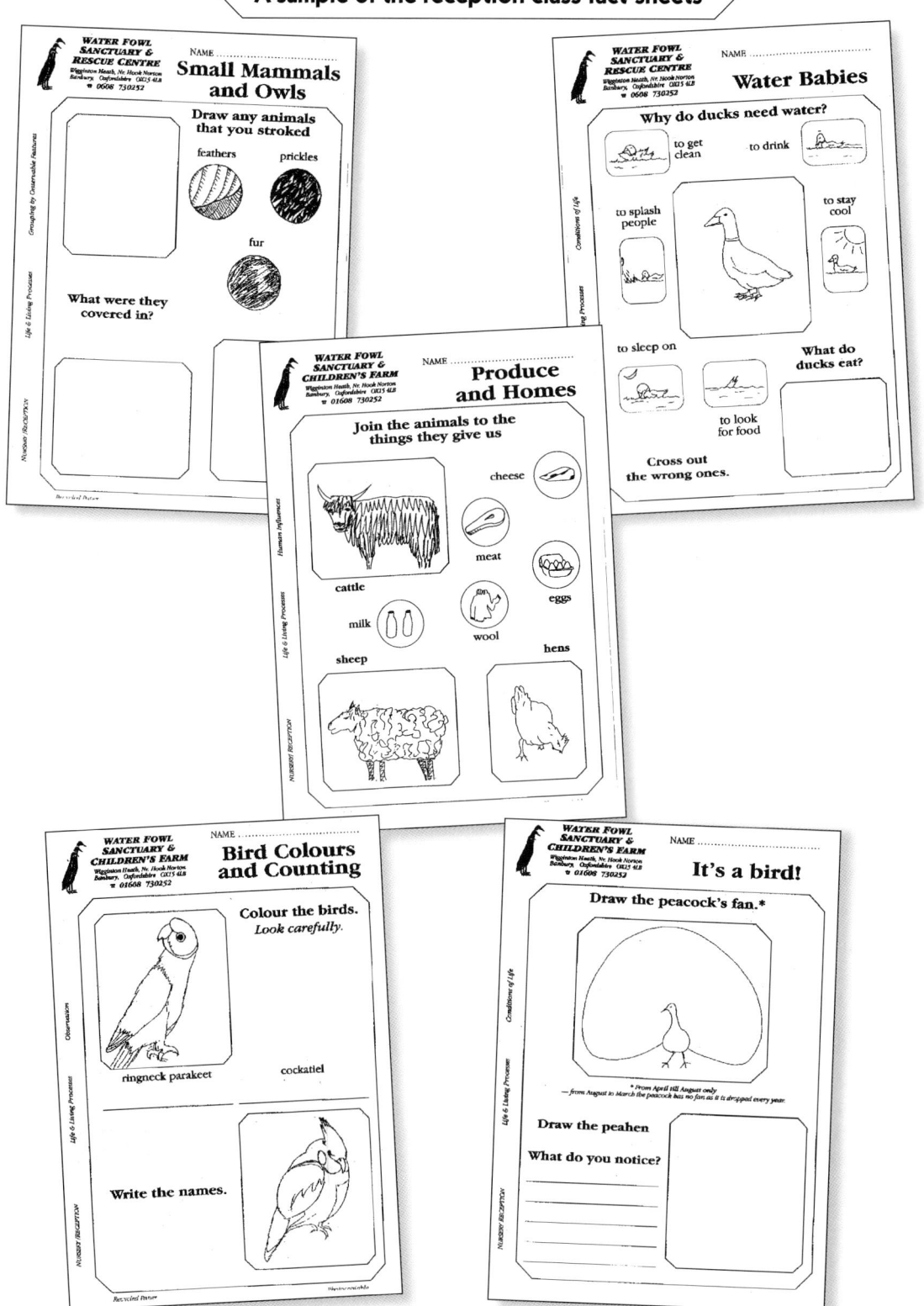

Our Michael brought some Barnevelda chicks which he has hatched.

MARCH 25TH 1995

I attended a sale at Enstone Airfield and bought a load of timber (3" x 1½" x 10' long) for £173 to make more aviaries.

I bought two pairs white Jap Bantams from Judy at Middle Barton.

MARCH 26TH 1995

Very busy with visitors.

MARCH 27TH 1995

I sent a letter to the council about planning for a bungalow on site. First Soay lamb born in the afternoon.

MARCH 28TH 1995

Wootton School, Woodstock, had an accident with one of my incubators, and so were not able to hatch any chicks. I received a letter of apology from them plus £20 for repairs.

MARCH 30TH 1995

Pat Lakin and Heather from Kingham brought Silkie Cocks; they are going to get other colours from Bedford for us all to breed — kind ladies.

MARCH 31ST 1995

Cousin Audrey and Don from Wales visited; they brought Auntie Agnes, 90 years old, my Dad's beloved sister, and also Joyce and Gordon. It was lovely to see Audrey and Don after so many years.

MARCH 30TH 1995

One of our tawny owls is ill.

Paul, a young policeman, brought in another Tawny owl this evening after contacting the Owl and Hawk Society who told him: *"Oh, he'll die in three hours if he's been hit by a car."* They didn't want to know on a Friday evening. They were closed until Monday morning.

APRIL 2ND 1995

Overflowing with visitors.

The RSPCA did not want to help a barn owl brought in today. A caring lady who had seen the bird on the side of the road took the trouble to turn round her car, stop, and pick it up before getting in touch with me. She brought it all the way from Newbury. It was starved almost to death, and had obviously been hit by a vehicle, not having the strength to fly out of the way.

APRIL 3RD 1995

Bletchington School came and loved it. *See letter pages.*

APRIL 4TH 1995

Cuttleslowe School visited plus Fairway Nursery. Again a pleasure!

APRIL 5TH 1995

Visit from Chadlington School. *See letter pages.*

Barn Owl

APRIL 6TH 1995

Pat Lakin brought eight blue Silkies and had travelled 120 miles to buy them for Heather, herself and me. They are show quality and very nice.

Broughton is all you expect of a stately pad

By PAUL WARNER

■ Lord of the manor, Lord Saye and Sele with the castle in the background – not so much a stately building but his home.

FACT FILE

OPEN: Broughton Castle is open every Wednesday and Sunday until September 14. It is also open every Thursday during July and August and Bank Holiday Monday.
TIMES: Opening times are from 2-5pm.
ADMISSION: £3.50 for adults, £3 for senior citizens and students and £1.50 for children.

A 'treasure house' of sheer delight

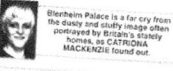

Blenheim Palace is a far cry from the dusty and stuffy image often portrayed by Britain's stately homes, as CATRIONA MACKENZIE found out.

■ ENGLISH SPLENDOUR: The formal water gardens as seen from Blenheim Palace. Acres of sweeping grounds include woodland, pleasure gardens and a maze.

FACT FILE

Banbury GUARDIAN Plus

Sanctuary is an ideal choice for those close encounters

ANYONE with a love of animals will find much to please them at the Wigginton Waterfowl Sanctuary and Children's Farm. Situated a few miles from Banbury, the sanctuary is a haven for thousands of birds and animals and doubles as both a tourist attraction and an animal rescue centre. SIMON ATTWOOD visited the sanctuary this week to find out what it has to offer people looking for a day out.

IT'S always refreshing to visit a tourist attraction that's been created with love rather than by a desire to make bucketloads of cash.
Such a place is the family-run Wigginton Waterfowl Sanctuary and Children's Farm which not only attracts thousands of visitors every year but also makes an outstanding contribution to the environment by looking after sick or unwanted animals.
Anyone, even those with just a passing interest in animals, will find much to fascinate them.
Not only are there hundreds and hundreds of ducks, geese and swans to look at, but there's a plethora of other animals including rabbits, pigs, chipmunks, goats, sheep and unusual creatures to which I couldn't even put a name.
Perhaps the most spectacular sight at the sanctuary is the magnificent peacocks, all of which are keen to spread their feathers and strut their stuff at the merest hint of a female.
The sight of these spectacular birds close up more than makes up for the entrance fee.
The real beauty of the sanctuary is that you can get as close as you like to the animals. In fact, owner Mabel Warner actively encourages you to pick up and stroke the animals.
"This sanctuary is for children of all ages," she said.
"Children are encouraged to handle the chicks, ducklings and baby rabbits.
"Not only does this make them more tame but it also helps to teach the children how to be gentle and kind."
Now you may think that looking at a lot of ducks and geese would get a tad tiring after a while, but you'd be wrong. They are fascinating creatures, especially the males who seem to spend much of their time trying to act butch by sticking their chests out, walking round with a superior look on their faces and squawking a lot.
Now is a good time to visit the sanctuary for spring has well and truly sprung and so have lots of little chicks who are breaking out of their eggs at alarming speed in the sanctuary incubator.
"We have eggs hatching for most of the year but now is obviously a pretty busy time with much to see," said Mabel.
The sanctuary, which also caters for school groups, is spread over 22 acres with ponds, streams and enclosures providing a safe and natural environment for the birds.
She said: "The site has never been treated with herbicides or pesticides.
"When wild birds are released after convalescence, they find a balanced, natural diet on hand and often remain in the vicinity."

■ Peacocks give you a spectacular welcome to the sanctuary.

WIGGINTON WATERFOWL SANCTUARY
This voucher will admit **two** children to the sanctuary free when accompanied by **two** adults.
GUARDIAN 1995 ONLY

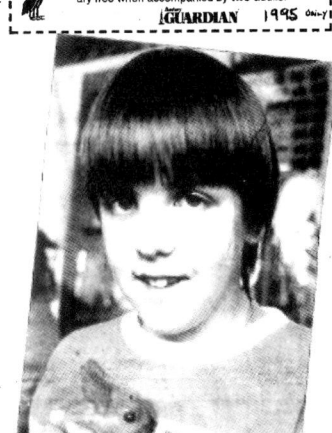

■ Eleven-year-old Laura Green finds a new friend.

FACT FILE

WIGGINTON Waterfowl Sanctuary and Children's Farm is roughly halfway between Banbury and Chipping Norton - just follow the brown duck signs from the A361 Banbury to Chipping Norton road or the B4035 Banbury to Shipston road.
Admission Prices:
Adults: £2.50
OAPs: £2.00
Children: £1.50
Special rates for large groups and coach parties. Ring 01608 730252 for details.
Open every day from 10.30am to 6pm

■ IT'S feeding time for these water fowl at the sanctuary – and anyone can join in the fun, young and old.

Bloxham brownies visited this afternoon and loved every minute. Late opening appreciated.

APRIL 8TH 1995

I am one year older today, but feeling younger!

The tawny owl brought in by the policeman on Friday has been released into a thirty-foot flight pen as he is now eating after seven days of refusing feed of any kind. It is a miracle he is still alive.

At last the barn owl is also recovering well and on the way to freedom.

APRIL 9TH 1995

A busy day with lots of lovely visitors again.

APRIL 10TH 1995

The two pet lambs that came from Anne and Tom Sammons are great favourites with everyone. Sara, the tiny Jacob sheep, gave birth to an even tinier black lamb. The other two Soay lambs are strong and lively, a male and a female; they are in the new small pens above the adventure playground with their mother.

One visitor said she liked my little notices about the place. There is one in the loos saying *"Sorry, hope you don't mind sharing the loos with the birds."* If I forget to close the doors, the pigeons fly in at night and roost on the ledges and we all know what pigeons do, don't we? There is another notice by the cuddling ash trees. In 1989, my husband was going to cut one ash away from the other as they had grown entwined around each other. I said *"No, no, no"*. I had an immediate inspiration about what to do with the trees. I asked Roy to make a seat for two below them and wrote:— *"Lovers sit for a while beneath our cuddling ash trees and count your blessings."* As they are growing at the top end of the Sanctuary it is a very relaxing spot, with a lovely view looking down at all the ducks enjoying themselves in the ponds, busily foraging about for worms and being fed by visitors.

There are other little notes about birds and animals living together in the cages inside the barn area. The parakeet dearly loves the rabbit he lives with and they kiss frequently. I bought him a mate but he fought her off, so she had to be moved next door. He now has a choice of two wives, but still prefers his Ruby Cashmere lop-eared rabbit, one of the biggest rabbits I have ever seen. She has beautiful white long fluffy fur. When I get her out of the cage to clean her underparts, Cocky, the parakeet, comes out, too, and sits beside her, then flies back in afterwards. They are inseparable.

APRIL 11TH 1995

I took Snowy another lop-eared Cashmere mother rabbit into the West Bar Veterinary Hospital for the removal of a lump which had appeared underneath her body. I hope it is not malignant, as she is such a good mother. Her treatment cost £52.

APRIL 12TH 1995

John, the travelling baker from Banbury, kindly leaves stale bread with Pam, who works at the Service Station in Bloxham. The bread is used to help feed the rescued birds and animals. Pam, from Milcombe, is a good friend and brings the boxes up to the Sanctuary, sometimes accompanied by Brian, her husband. *(See colour plate.)* There is a long list of well-wishers bringing in stale or unwanted bread — among them John and Chrissy from New Road Stores, Milcombe; John & Kathleen, Adam Stores at Enstone; Anne from Hook Norton Post Office & Stores; and Doreen Roberts from Wigginton. *(See colour plate.)* All help me feed the ever-growing family and I am very grateful to them all. Charlie Ward from Shutford kindly buys green cabbage, lettuce, celery and carrots in memory of his beloved late wife, both he and she having shared a great love of rabbits.

Disneyland

Joy from Horspath brings used feed-bowls with her granddaughter, Sally, who says the Sanctuary is her favourite place. On one occasion Sally had the opportunity to go to *Disneyland* near Paris, but preferred to come here. *See letter page 150.*

APRIL 14TH 1995

Good Friday. We are six years old today and the car park is full. Tony and Pippa Jacobs from Didcot brought ten children: James, Barney, Emily, Thomas, Nicholas, Dorothy, Tabitha, Rosemary, Lucy and Benjamin. The mother said: *"this is a unique place, so safe — I can really relax and let all the children enjoy themselves."* They were all very well-behaved and beautifully brought up, the kind of children I adore.

Derek, our postman from South Newington, brought in a Sussex Cockerel and two Buff Orpingtons. They had been destined for dinner, but were just too beautiful to kill.

APRIL 26TH 1995

Jane Gillett came with some of the first typing for this book.

MAY 10TH 1995

A school party came from Standlake with their Headteacher, Sue Harrap — very happy. They were all very well-mannered little children, most of them said *"thank you"* as they left.

Late this afternoon a couple came just to sit among my babies, as sadly they had lost their baby son prematurely. My place is the only place the young wife wanted to be. They stayed for about an hour, she just sitting in her own chair among my babies. I let her hold the tiny ones just born, and she asked if any had been born on May 1st, the day her little one died. I was able to take from the nest a handful of wriggling little bodies just opening their eyes, ten days old. She was so grateful and what devotion her young husband showed to his treasured young wife. It did me good to see such tenderness in these days of cruelty and anger. They said they had visited earlier in the year and enjoyed it very much. She told me she was a teacher at Blackbird Leys in Oxford, but I forgot to ask their names.

Thank you my duck!

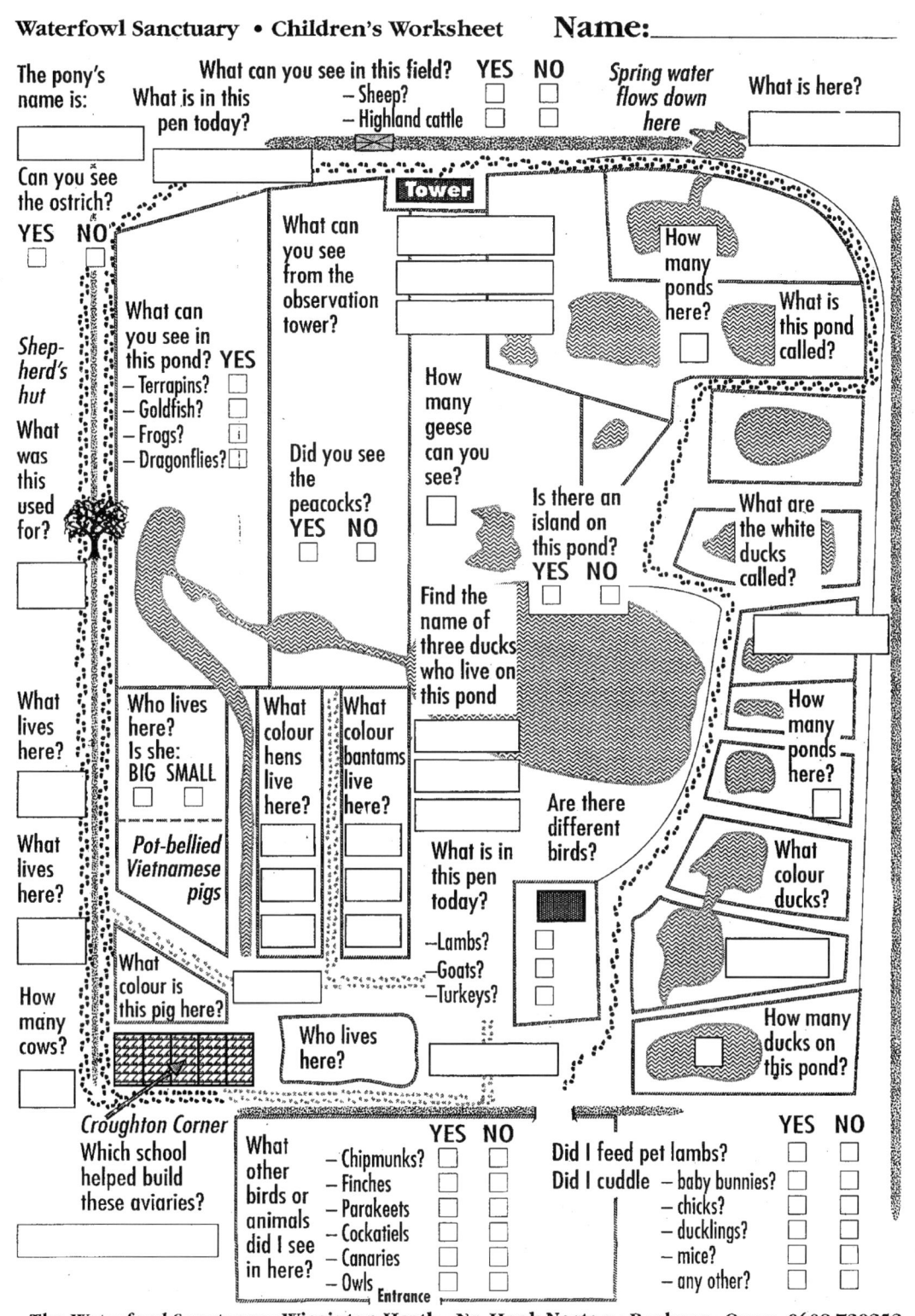

Waterfowl Sanctuary • Children's Worksheet   Name:_____

The pony's name is: _____

What is in this pen today? _____

What can you see in this field?
- Sheep? YES ☐ NO ☐
- Highland cattle? YES ☐ NO ☐

Spring water flows down here

What is here? _____

Can you see the ostrich? YES ☐ NO ☐

Tower

What can you see from the observation tower? _____ _____

How many ponds here? _____

What is this pond called? _____

Shepherd's hut

What was this used for? _____

What can you see in this pond?
- Terrapins? YES ☐
- Goldfish? ☐
- Frogs? ☐
- Dragonflies? ☐

Did you see the peacocks? YES ☐ NO ☐

How many geese can you see? _____

Is there an island on this pond? YES ☐ NO ☐

What are the white ducks called? _____

Find the name of three ducks who live on this pond _____

What lives here? _____

Who lives here? Is she: BIG ☐ SMALL ☐

Pot-bellied Vietnamese pigs

What colour hens live here? _____

What colour bantams live here? _____

Are there different birds?

How many ponds here? _____

What lives here? _____

What is in this pen today?
- Lambs? ☐
- Goats? ☐
- Turkeys? ☐

What colour ducks? _____

What colour is this pig here? _____

Who lives here? _____

How many cows? _____

How many ducks on this pond? _____

Croughton Corner

Which school helped build these aviaries? _____

What other birds or animals did I see in here?
	YES	NO
Chipmunks?	☐	☐
Finches	☐	☐
Parakeets	☐	☐
Cockatiels	☐	☐
Canaries	☐	☐
Owls	☐	☐

Entrance

Did I feed pet lambs? YES ☐ NO ☐

Did I cuddle
- baby bunnies? YES ☐ NO ☐
- chicks? ☐ ☐
- ducklings? ☐ ☐
- mice? ☐ ☐
- any other? ☐ ☐

The Waterfowl Sanctuary · Wigginton Heath · Nr. Hook Norton · Banbury · Oxon. 0608 730252

Our most popular Children's sheet to date

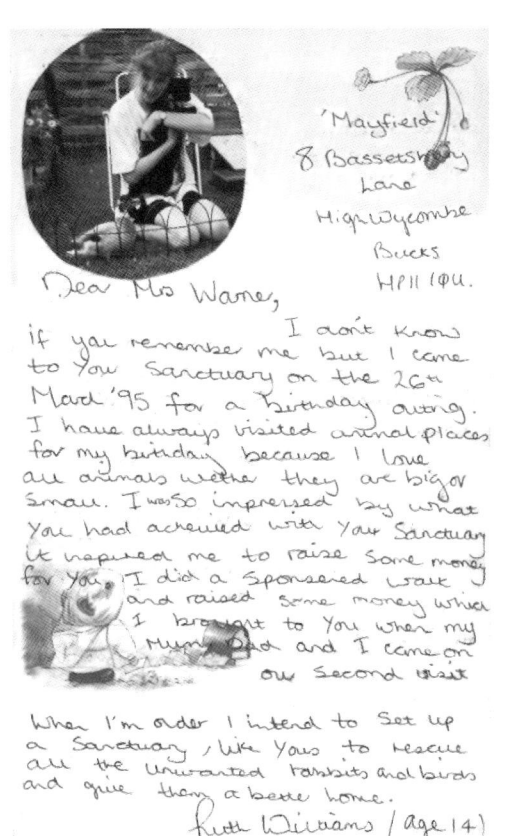

> Dear Mrs Warne,
>
> 'Mayfield'
> 8 Bassetsbury Lane
> High Wycombe
> Bucks
> HP11 1QU.
>
> I don't know if you remember me but I came to your Sanctuary on the 26th March '95 for a birthday outing. I have always visited animal places for my birthday because I love all animals whether they are big or small. I was so impressed by what you had achieved with your Sanctuary it inspired me to raise some money for you. I did a sponsored walk and raised some money which I brought to you when my Mum, Dad and I came on our second visit.
>
> When I'm older I intend to set up a Sanctuary, like yours to rescue all the unwanted rabbits and birds and give them a better home.
>
> Ruth Williams (age 14) + animals

MAY 20TH 1995

A little girl from High Wycombe loved my place so much she went home and organised a sponsored walk, raising £15. She walked for two and half hours and said she wanted to do something like this when she grows up.

MAY 24TH 1995

Martin and Rodney are building the extension on the side of large barn, 10 ft wide, to house babies, mainly for winter comfort and it will make the children's farm much more interesting all year round.

JUNE 6TH 1995

Three coach loads of children came, two from Berkhamstead, one from Stadhampton: ninety children in all. They (and I) loved every minute of it.

JUNE 8TH 1995

Fifty-four children came from Birmingham — the usual ecstatic comments.

JUNE 10TH 1995

Maurice's surprise birthday party at Murcott Village Hall, – a lovely gathering of relatives and friends.

With her mother Rosie, little Georgina Hill came with a card she'd made for me. It was her birthday and she wanted to come here rather than having anything else.

A mini-bus came from London, another birthday party return visit. They came first at half-term, then returned today with fourteen little friends. Twenty-five parents and children from the Witney area also came to hold a birthday party. The weather was beautiful.

I met Mr. House who said St. Francis will be waiting on the steps for me as I go up out of this earthly classroom! He's 83 years old.

JUNE 23RD 1995

Today Wendy and Michael had their wedding at Wigginton Chapel – a small gathering of about thirty friends and relatives. Richard V., who made the video for the Sanctuary this year, took the photographs. Melody was bridesmaid and Micah, page boy. Helen, wife of Harry of Tudor Photography, Banbury, stood in for me at the Sanctuary. She often brings little Ben and Hazel here as, in the early days, Harry kindly took several professional photos for me, some of which are used in this book, (no charge).

JUNE 24TH 1995

My friend Ann Morton, head-teacher at Finmere School, and her husband, David, brought back yet another incubator. When they are returned from the schools he is very kindly overhauling them for me free of charge.

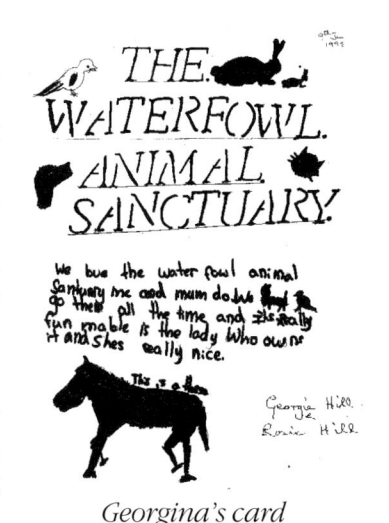

> THE WATERFOWL ANIMAL SANCTUARY
>
> We bus the waterfowl animal sanctuary me and mum do lots of their all the time and its really fun mable is the lady who owns it and shes really nice.
>
> Georgie Hill
> Rosie Hill

Georgina's card

Building of the Croughton Corner 1993

More heavy work for Rodney *The completed aviaries*

Brian, a Bloxham policeman came and told me there was lots of useful timber being thrown into skips opposite Banbury police station, so after hasty phonecalls to the foreman of the site, we were permitted to remove several van-loads out to the Sanctuary. Much of this was used in the construction of the doors both to the Croughton Corner and the Baby Barn extension.

Building of the Baby Barn Extension 1995

Martin Cherry and Rodney laid the concrete floor, and Madge and Colin Stockford from Swerford built up all the concrete blockwork for the lean-to barn extension

Martin (left) and Rodney (right) lay the concrete floor for the extension

Madge & Colin *"995" our faithful 25-yr-old David Brown*

I received a copy of a letter from David Shepherd, our caring vet from West Bar Veterinary Surgery, who is backing my attempt to build a stockman's bungalow at the Sanctuary. Additionally Marcus Hughes is backing me with a letter to the council.

June 30th 1995

David Price visited to inspect the Sanctuary prior to the arrival of thirty six Greylag geese, and an assortment of domestic ducks. He and his colleagues from Slimbridge are bringing them from London, as they are proving a nuisance by over-grazing the grass. I must make a further pen in the top field. I received an estimate for further pens there from Parkland Fencing – but the project is temporarily shelved!

July 1995

Rodney is off on his travels again.

July 3rd 1995

Chaps from Slimbridge arrived with thirty Greylag Geese, four Muscovy ducks, and two young ducklings: all very nice and tame.

July 5th 1995

Mr. and Mrs. Shell from Hendon, London, came with guinea pigs which they breed in a cellar, three boxes full, just as they have done in the past three years The trouble is, they won't bring them young. They are a childless couple and do not appreciate the value of little children handling them. I find this a little sad as I could sell them for Sanctuary funds if they were brought in young, but I end up giving them all away from the end of October. I cannot keep them through the winter as it is too cold and wet at Wigginton.

A few more baby ducklings arrived from Tina and Colin Chester, who have been hatching a few ducklings for me each year at Upfield, Greatworth.

Thirteen more guinea pigs donated, young ones this time, with shiny little bodies, which can be sold at £8 each.

Kay from Coventry rang to say she has about forty more white doves to bring soon.

A family from Finland came and loved the place.

July 12th 1995

Today schoolchildren came from Chatsworth Primary, Hounslow, Middlesex; also children from Longfield, Bicester, with Mrs. Joan Warren in charge. A playgroup from Lois Weedon arrived, then at 4.30 p.m. the Pets' Club from Henry Box School, Witney, came. The club have been five times and learn a lot at the Sanctuary with Bob Morris.

Jenny rang to say she had lost one of her special runner ducks, her son's favourite, and asked whether I had a replacement, which luckily I did.

July 14th 1995

Cousin Terry Saxton from Botley brought in rabbits, either to sell or to find good homes. I do find new homes for most of them, but big baby rabbits are unsaleable, I find.

July 1995

Richard V. finished painting the barn lean-to in a lovely dark green shade – all voluntary work.

My mission in the later years of my life in the 1990s is to try to educate children of all ages to show respect, love and care for all God's creatures as they are a very important aspect of our life in Britain. I hope the work I have done has to a small degree achieved this aim.

Another two big rabbits re-housed.

Yet another terrapin has been brought in. The lady released it into the pond herself and left £10 for Sanctuary funds; another £10 came from Kate Woolley for taking in her beloved cockerel, Bruno.

JULY 25TH 1995

The three lads from Slimbridge brought another six ducks and geese from London with a promise to help make new pens and care for the birds.

Two different large rabbits have once again been dumped here, just pushed through the gate. If folk won't leave money for vaccination against myxomatosis, then I am terribly afraid it will strike the rabbits down. Many varieties of thistles growing around the Sanctuary are believed by some people to harbour fleas which carry the dreaded disease. It is a horrible death for poor rabbits. The summer months are the usual time when all my stock are vaccinated against it – followed, of course, by a large vet bill.

JULY 26TH 1995

Martin and John Cherry combined Den's eight acres of barley, a lovely golden harvest, bright golden bales of straw, a hundred bales were stacked in the barn with Rodney's help.

Four young blue peacocks hatched. All the adult male blues are shedding their beautiful long tail feathers now and I have an array on my counter. Peacocks shed their whole tails from June to August then grow them again for the following spring display / pairing. I'm afraid I have to confiscate the ones picked up by visitors or invite them to pay £1 each, as they are our harvest and can be sold throughout the year.

Alan and Shirley Boswell from Bloxham brought children, Anna, Danielle, Jamie and Danyka to choose a pet rabbit. When they had chosen Alan said *"We'll take this one my duck"*, the typical Oxfordshire phrase which I have used for the title of my book.

The springs have stopped running and the water gets quite green when it stops flowing, but ducks and geese all seem very happy. Most of the ponds are good and deep since we have had them cleaned out.

"We'll take this one, my duck!"

AUGUST 1ST 1995

When I applied to Cherwell District Council to build a stockman's bungalow at the Sanctuary my application was deferred on the grounds that I do not make enough profit — although it is now definitely improving. I was up most of the night writing letters inviting support of my case from representatives of several institutions, including the chairman of the National Farmers Union; and the Chief Officer of the Wetlands and Wildfowl Trust, Slimbridge, David Price, (who recently organised the delivery here of the Greylag geese from London.)

AUGUST 4TH 1995

A coach-load of schoolchildren came from Birmingham and another from Cuttleslowe, Oxford, and also three small coach-ambulances carrying children in wheelchairs from the Kent Adventure Club. All seemed to enjoy themselves.

Richard Payne has re-written the letters to all councillors, and also a reply to John Simms, the architect, re planning. I will be sending videos to all of them.

The Chinese Goose arrived at Tresco, Isles of Scilly, in good shape. Dr. Ron requested a grey female Chinese Goose, which he couldn't find anywhere else. I was able to supply one to join the remaining flock of males.

Chinese Goose to save the flock on Tresco

In 1863 a flock of extremely rare grey Chinese Geese was rescued from the wreck of the "Friar Tuck" when it sank off St. Mary's Island, Tresco, in the Scilly Isles, during the worst storm in living memory en route to Europe from China. The geese have prospered there for well over a century, but were threatened with extinction due to the fact that there were no remaining females. Through the British Waterfowl Association, Dr. Ron Gleadle of the Tresco Estate heard of my work and enquired if I could supply a young female of the species to preserve this historic flock. He also contacted a friend, Chris Hawes, of Banbury. Happily I was able to arrange for the Estate to receive one of my two females, and Chris transported her safely down to Tresco during this very hot summer of 1995. Preserving rare breeds in this way is another important aspect of the life of the Sanctuary.

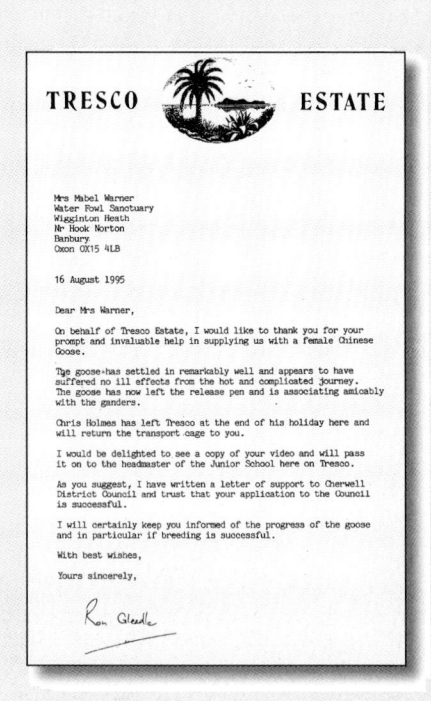

AUGUST 8TH 1995

I went to Cherwell District Council offices. Planning committee has decided to defer my planning application until September 7th, when I hope to be able to prove beyond doubt that mine is a viable business and capable of employing one person full-time. They say they have to view it long term say, over a hundred years. What will it be like then? I sent off fifteen videos of the Sanctuary to all councillors who deal with planning applications.

AUGUST 14TH 1995

Cousin Terry Saxton from Botley, Oxford, helped me for an hour or so, repairing fences in the chicken runs, putting new wire along the bottom where it has rotted away over six years — a good voluntary job. Thank you, Terry. Richard V. has been clearing up the barn again, voluntarily. All good helpers.

No rain, and bright sunny days bring lots of visitors, mostly grannies with little children; also quite a few foreign visitors. Baby animals have no language barriers; they are loved by children of all nationalities.

June and Terry

AUGUST 25TH 1995

Copies of the supportive letters from Stan Bowes of the Southern Tourist Board and Dr. Ron of Tresco. However, following a meeting with Ian Grace, chief planning officer with the northern area, Cherwell District Council, I saw a copy of a letter from Wigginton Parish Council recommending refusal on grounds of suspected contravention of the former planning conditions. They say, (they suspect, mind), I have not abided by the conditions of the mobile home being used as storage only. This is not true as Wendy moved along the road next door to the Manor Farm Cottages, then went to live at South Newington with her new husband.

SEPTEMBER 7TH 1995

Planning meeting — seven for the application and seven against, so it was deferred until the next meeting. At least it was not overwhelmingly one-sided. There was surprise amongst the councillors that the Sanctuary has lasted this long, – six and half years. Some only gave it six months. I have four more weeks to try and persuade them.

Cyril's Mum

SEPTEMBER 9TH 1995

I went with Den to Stoneleigh Rare Breeds show/sale. Den bought a trio of Light Sussex hens. Richard V. and Terry manned the Reception voluntarily.

SEPTEMBER 12TH 1995

Their interest having been aroused, a reporter from *Radio Oxford* came specifically to tape an interview with me about the planning. It was broadcast this afternoon at 2.10p.m. and lasted for a quarter of an hour. In it I emphasised the importance of the bungalow being built on-site, and overall we gained some sympathetic publicity which should be very beneficial to our cause.

Interviewed by Radio Oxford

SEPTEMBER 14TH 1995

A local councillor is dead set against the Sanctuary building a bungalow on the site. He wanted further proof that it is a viable venture and a business, not a hobby. Some councillors are not the slightest bit interested in the rare breeds I nurture here. All they are interested in is cold calculation, — money, money, money. Visitors and children's happiness matters not; education of the natural environment matters not; care and love matter not. Indeed, as the letter from Councillor Hawthorn, of Birmingham, shows, no relief is acceptable in this cruel world we live in – just greed and selfishness. All the eight years' work and dedication I have put into the Sanctuary counts for nothing. Only one councillor has been out to see the site.

SEPTEMBER 15TH 1995

The local councillor visited again. It was a much more pleasant encounter and I gave him more letters of support to read when he had got over his jet-lag, (he had just returned from abroad.) I hope he comes around to seeing sense when he knows how much work has gone into the Sanctuary over the last eight years. He had no knowledge of the planning application until one hour before the meeting.

SEPTEMBER 18TH 1995

A good supportive letter from Geoffrey Pearce, of *Brinsea Products* (incubators). I'm posting off more copy letters to the fifteen planning officers: I hope I don't overdo it. I feel they need to know how desperately I need this residence at the Sanctuary.

SEPTEMBER 19TH 1995

A copy of another support letter from Mr. Johnson of the N.F.U.

Colin Turner, my friendly bank manager, visited and agreed to finance the bungalow.

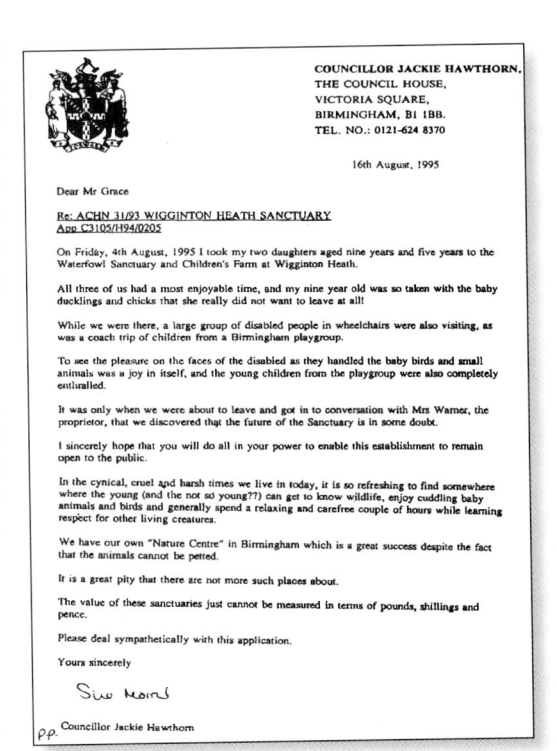

COUNCILLOR JACKIE HAWTHORN,
THE COUNCIL HOUSE,
VICTORIA SQUARE,
BIRMINGHAM, B1 1BB.
TEL. NO.: 0121-624 8370

16th August, 1995

Dear Mr Grace

Re: ACHN 31/93 WIGGINTON HEATH SANCTUARY
App C3105/H94/0205

On Friday, 4th August, 1995 I took my two daughters aged nine years and five years to the Waterfowl Sanctuary and Children's Farm at Wigginton Heath.

All three of us had a most enjoyable time, and my nine year old was so taken with the baby ducklings and chicks that she really did not want to leave at all!

While we were there, a large group of disabled people in wheelchairs were also visiting, as was a coach trip of children from a Birmingham playgroup.

To see the pleasure on the faces of the disabled as they handled the baby birds and small animals was a joy in itself, and the young children from the playgroup were also completely enthralled.

It was only when we were about to leave and got in to conversation with Mrs Warner, the proprietor, that we discovered that the future of the Sanctuary is in some doubt.

I sincerely hope that you will do all in your power to enable this establishment to remain open to the public.

In the cynical, cruel and harsh times we live in today, it is so refreshing to find somewhere where the young (and the not so young??) can get to know wildlife, enjoy cuddling baby animals and birds and generally spend a relaxing and carefree couple of hours while learning respect for other living creatures.

We have our own "Nature Centre" in Birmingham which is a great success despite the fact that the animals cannot be petted.

It is a great pity that there are not more such places about.

The value of these sanctuaries just cannot be measured in terms of pounds, shillings and pence.

Please deal sympathetically with this application.

Yours sincerely

Sue Moins

p.p. Councillor Jackie Hawthorn

Thank you my duck!

Cotswold Farm Park
RARE BREEDS SURVIVAL CENTRE

JLH/AES

21st September, 1995.

I. Grace, Esq.,
Chief Planning Officer 'North Area),
Cherwell District Council,
Bodicote, Banbury, OX15 4AA.

Dear Mr. Grace,

I have been in contact with Mrs. Warner of the Water Fowl Sanctuary and Children's Farm at Wiggington Heath, Banbury who tells me that she is applying for planning permission to build a stockman's cottage on the site in order to make it possible for her to continue her work and remain open.

Mrs. Warner does an important job of conservation and education introducing visitors of all ages to the importance of preserving our living heritage, wildlife and countryside.

To have a member of staff living on site is absolutely essential from an animal welfare point of view. The night is the most important time for attention to birds and animals. A very high percentage of births seem to take place during the night and it is at night that they are at greatest risk from predators such as foxes, badgers, stoats, weasels and mink and sadly, in this day and age, human vandals.

I strongly urge you to support this application for a dwelling. A friend of mine who has a Farm Park in Devon, and does not live on site, came to work a few weeks ago to find that vandals had released all his poultry and rabbits and allowed terriers to rip them to pieces. Apart from the horror of finding it, you can imagine the headlines in the local press.

Yours, with serious concern,

Joe Henson

J.L. Henson.

Founder Chairman Rare Breeds Survival Trust
and a past Vice President.

The Domestic Fowl Trust
Honeybourne, Evesham, Worcestershire, WR11 5QJ
Registered No. 1962487
V.A.T. No. 286 1303 63

Telephone: Evesham (01386) 833083   Fax: (01386) 833364

Mr Ian Grace
Chief Planning Officer
(North) Cherwell District Council
Oxfordshire

Dear Sir

Ref. Planning Application IG.950095/F for a stockman's dwelling at the Waterfowl Sanctuary and Children's farm at Wiggington Heath.

I am writing to you to support the application that Mrs Mabel Warner has lodged with you. I have known Mrs Warner for many years, and know the hard work and dedication she has put into the waterfowl sanctuary.

The importance of having a dwelling on site is paramount. I know this from first hand experience, as only this year some lunatic cut our perimeter fence in order to allow foxes to gain entry, and had I not been living on site, the damage would have been enormous. This is one of the main headaches of keeping livestock and being open to the public. There is always going to be some person, who because of some reason, will break into a place not only to steal but also to maim or harm the stock. This factor is most important when considering an application such as Mrs Warner's.

If you wish to contact me I would be only too delighted to speak to you.

Yours sincerely

Michael Roberts — I hope this does the trick

Michael Roberts

Jeddah Preparatory School
British Consulate General
P.O. Box 6316
Jeddah 21442
Saudi Arabia

Mr Ian Grace
Planning Officer
Cherwell District Council
Bodicote House
Banbury
Oxon.

11 July 1995

Dear Mr Grace,

Water Fowl Sanctuary & Children's Farm, Wigginton Heath.

Because of contractual obligations in the Middle East, my family only live in our house in Banbury during July and August every year.

My three daughters, all aged below six years, particularly enjoy investigating and participating in the type of activities in the Banbury area that are ordinarily denied to them in the Middle East.

The various parks, Spiceball Centre, the Mill, the local summer fairs to name but a few, all serve to enrich our short stay in the area. However one institution is consistently the firm favourite with the children.

Our brief stay is punctuated with at least half a dozen visits to the Water Fowl Sanctuary & Children's Farm in Wigginton Heath. The opportunity to touch and see so many varieties of birds kept in a most charming, informal environment is a credit to the management. I sincerely hope that they can rely upon your generous support in the future.

Yours sincerely,

Mr Alan R. Jones.
6 Calthorpe Road
Banbury
Oxon OX16 8HS.

PARAGON PRINT & DESIGN

THE OLD CHAPEL,
ARNSBY COTTAGE,
MILCOMBE,
BANBURY,
OXON. OX15 4RP

Tel & Fax: 01295 721370

October 3rd 1995

Mr Ian Grace,
Chief Planning Officer (North Area),
Cherwell District Council,
Bodicote House, Bodicote,
Banbury, Oxon. OX15 4AA

Application no. 95/00495/F. – Sub-Committee meeting: October 5th
for a stockman's cottage at The Waterfowl Sanctuary & Children's Farm

Dear Mr. Grace,

In connection with the above application, I would like to support the views of those who have already written to you urging the Council to allow the building of one cottage at the Sanctuary. I have been glad to supply and help Mrs. Mabel Warner with her advertising, literature and stationery over several years and have witnessed how the Sanctuary has grown to become a viable business and much-loved local tourist attraction.

I am, of course, concerned that Mrs. Warner might feel it necessary to close the Sanctuary if this permission is not granted, and I need not tell you that this would have a considerably bad effect on my own business, too.

The possible effect on the local community, however, would be devastating. Those who have been to the Sanctuary (from near and far) and have seen what is available, could not in all honesty oppose this application for a stockman's cottage. If they were able to see and understand the reasons, the value of the work done there, both for species survival and for the education of children and families then the decision to grant permission would be compelling. There would be no hesitation.

In the course of my work I have come to see the Sanctuary as a rare jewel rather than a trinket, precious rather than expendable. The children of Milcombe, (my own included), have come to love the Sanctuary more and more; there they can have true hands-on experience of birds and animals which is not readily available elsewhere, and learn in an enjoyable way. Qualities of gentleness, kindness are caring are instilled in what is, in many ways, a hard and ruthless world.

If the Sanctuary fails to fit exactly into Local Plan categories, it is because it is an exceptional place and Mrs. Warner and her son have nothing but good and honourable intentions.

I apologise that this letter has come to you rather late, but I hope, at least, it is not too late for you to make the Council aware of its content.

Thank you for your attention,

Yours sincerely,

RMAPayne

Richard M. A. Payne, B.A.(Oxon.)

Letter 1: Southern Tourist Board

40 Charterhouse Road, Eastleigh
Hampshire SO53 3JH
Tel: National – 01703 620006
Fax: National – 01703 620010
Tel: International +44 1703 620006
Fax: International +44 1703 620010

President
The Lord Montagu of Beaulieu

Mr Ian Grace
Chief Planning Officer
(Northern Area)
Cherwell District Council
Bodicote
BANBURY
OX15 4AA

20th August 1995

Reference: Planning Application No. 95/00495/F

Dear Sir

I write in reference to the above planning application for the provision of a suitable dwelling on the site of Wigginton Waterfowl Sanctuary.

Mabel Warner has been known to me ever since this sanctuary was first established and through her hard work and commitment it has become a well established Tourist Attraction (and much needed) in the area.

The need for a permanent dwelling after five years of operation is due in part to the need to have a more effective form of security for the very essence of the attraction - the livestock! It is an unfortunate that in the Tourism industry related to tourist attractions that the theft of livestock is now a real hazard with which the owners have to contend - Refer to The Cotswold Wildlife Park and the spate of thefts that have occurred.

I would be grateful if you would advise the Panning sub committee of this support by providing the building design is such that it merges sympathetically within the location.

Thank you
Yours sincerely

Stan Bowes
Area Manager
Southern Tourist Board

From: Area Office The Town Hall Market Place
WALLINGFORD Oxon OX10 0EG Tel: 01491 825844

Southern England covers the Counties of Berkshire, Buckinghamshire, Dorset, Hampshire, Isle of Wight, Oxfordshire and South Wiltshire.

Letter 2: Elizabeth Reece

Mrs. Mabel Warner
Water Fowl Sanctuary & Children's Farm
Wigginton Heath
nr. Hook Norton
Banbury, Oxon, OX15 4LB

716 Maddux Drive,
Daly City, Ca. 94015

September 21, 1995

Dear Mrs. Warner:

I returned just a few days ago from a visit to England. I do want to tell you that one of the highlights of the trip for my daughter, two grandchildren, and myself was our visit on September 8 to the Water Fowl Sanctuary which you run.

It was a thrill for my grandchildren (ages 11-1/2 and 8) to experience the close contact you provide children with baby rabbits and chicks. For myself, I was impressed by the diversity of animals and birds (goats, Soay sheep, pigs, ducks and geese of many breeds) that you care for with obvious devotion, and you are to be commended for the general good health of your charges. Such an undertaking demands hard work as well as dedication.

Keeping the sanctuary open 364 days a year, indeed, looking after animals day in and day out is not a task for the fainthearted. That you have been running this very special place for over six years is testimony to your dedication. I do hope the authorities will give favorable consideration to your application for the building of a house for your son, enabling you, with his help, to carry on your good work.

Thank you for a most memorable visit and my very best wishes to you for continued success.

Sincerely,

Elizabeth Reece (Mrs.)

Letter 3: Councillor Ransome-Wallis

Re: ACHN 31/93 Wigginton Heath Sanctuary
APP C3105/H94/0205

My colleague, Councillor Hawthorn informs me that the Wigginton Heath Sanctuary is under threat unless planning permission for a Stockman's Dwelling is approved.

Obviously, I do not know all of the details regarding this application, but at the risk of "sticking my nose in where it is not wanted" may I say that I would consider it to be a tragedy if this Sanctuary was forced to close as it provides such a unique and enjoyable experience for young and old alike.

Please think "green" when considering this application!

Thank you.

Yours sincerely

Sue Morris

Councillor Christine Ransome-Wallis

Letter 4: West Bar Veterinary Hospital

Associate Veterinary Surgeons: Mr D A Shepherd BVMS MRCVS
Mr R M Jackson-Taylor BVetMed MRCVS
Assistant Veterinary Surgeons: Mrs E A Adams BVM&S MRCVS
Mr R F Byrne B.Sc BVSc MRCVS
Mrs K Peckitt BVM&S PhD MRCVS
Mr A M Roberts MA VetMB MRCVS
Miss J Price BVMS MRCVS
Mrs R Poissant BVetMed MRCVS
Mr C P Castle MRCVS
Mrs A Appleton BVSc

Shop Manager: Mr C R Argyle MBE
Drug Sales Manager: Mrs C E Gurney
Laboratory Technicians: Miss A Lawrence HNC
Mrs R Brownstone BSc

Practice Managers: Mrs M R Popplewell BA
Mrs S Steven

Head Nurse: Mrs A Appleton

COMPUTER REF: GRAJUN95.LET
IBM/DAS/ELG/TCPRO/RECEP

Mr I Grace
Planning Officer
Cherwell District Council
Bodicote
Banbury
Oxon
OX15 0AA   Ref 95/00495/F

21st June 1995

Dear Mr Grace,

I understand that Mrs Warner, the proprietor of Wigginton Wildfowl Sanctuary has applied for planning permission to build a stockman's house on the premises. This practice has been involved in providing veterinary services to Mrs Warner since she opened the Sanctuary in 1989. In my opinion it is very important for the welfare of the animals that she should have someone present on the site full-time, for the following reasons:

a) In case of illness or injury to an animal during the night.

b) Many of the animals involved are breeding animals and may be giving birth (or laying eggs) during the night, and would need supervision.

c) In case of any attacks by predators eg foxes.

d) From a security point of view - to prevent human intruders causing harm to the animals, either from malicious vandalism or in an attempt to steal them.

All in all, in view of the large numbers of animals and birds involved in the sanctuary, it is greatly to Mrs Warner's credit that she has managed to cope until now without a proper residence on the site. I would urge you to grant planning permission for a residence without delay, as otherwise the welfare of the animals may eventually be put at risk.

Yours sincerely

Mr D A Shepherd BVMS MRCVS

Letter 5: Brinsea

14th September 1995

Brinsea Products Ltd
Station Road, Sandford
Avon BS19 5RA, England
Tel: (01934) 823039
Fax: (01934) 820250

Mr I Grace
Planning Officer
Cherwell District Council
Bodicote, Banbury
Oxon, OX15 0AA

Ref: 95/00495/F

Dear Mr Grace,

The Waterfowl Sanctuary and Rescue Centre at Wigginton Heath is an essential distributor for our products in Oxfordshire and Northamptonshire.

We understand from the proprietor, Mrs Warner that the future development and security of the Centre depend on a house for a stockman on the existing site.

Apart from her interest in rare and endangered species, Mrs Warner actively promotes and supplies our incubators to schools where children can hatch eggs as part of their essential rural studies.

As a small manufacturing Company providing employment in the South West, we are keen to encourage such an admirable venture to grow without constraint. We therefore urge you to view favourably the planning application for a house at this premises.

Yours sincerely,

Geoffrey P. Pearce,
Director.

SEPTEMBER 20TH 1995

More letters of support arrived: one from my bank manager; one from Michael Roberts of the Domestic Fowl Trust at Honeybourne; also Pat and Heather told me they had written to Ian Grace. Pat is retired, having been head-teacher for twenty-six years, and is now advisor to Oxfordshire Schools.

Two ladies visited from Japan and only one spoke English. They were accompanied by three children and a nanny, all speaking Japanese. How they loved all the baby rabbits, chicks and ducklings. There's no problem of communication when there are babies – they make almost everyone happy.

A lovely little letter was sent from the children at Orchard Close, Sibford, thanking the animals for being here.

SEPTEMBER 22ND 1995

Valerie and Steve brought seed, American style, hot stuff, with many chilli peppers for parakeets, cockatiels and love birds. The white chipmunks purchased from Cambridgeshire also love these mixtures. As they are nearing 'sell by' date they are free and every little helps in feeding my large family.

Three more rabbits brought in; I also sold three of my Mabel's Midgets, one going up to North Yorkshire. Myxomatosis is in the wild rabbits around the Sanctuary, but all my stock has been protected by vaccination. Kate, the vet, recently came by to vaccinate two more.

I found good homes on a lake for four more drakes; also three large rabbits have been re-homed.

SEPTEMBER 24TH 1995

The peacocks brought in recently appear to be Javanese Green as the new feathers growing now are definitely very dark green on top of the wings, whereas the Indian Blues are mottled browns on the shoulders. The other two are speckled white and with good diet are looking far healthier than when they arrived.

PLANNING PERMISSION GRANTED

OCTOBER 5TH 1995

Halleluia! Planning permission has been granted for a permanent bungalow on site. After months of letter-writing, phone-calls and sleepless nights, trying to convince the Council it is right to allow a dwelling, I have finally won them over; the vote was eleven for, one against with three abstentions. Now, the Sanctuary will go on for many, many years. When leaving the meeting in Bodicote, I met Richard, who had run from Milcombe to find out the result.

Mabel beats building ban to set up sanctuary home

A CONSERVATIONIST says she is delighted planning permission has been granted to build a bungalow at her wildlife sanctuary.

Mabel Warner, pictured above, was given permission to build a stone bungalow on land at the Wigginton Waterfowl Sanctuary by Cherwell District Council.

Mrs Warner said: "I have been working towards this decision for the last seven years. The stock is increasing in number all the time. I need to protect the stock and ensure the survival of rare breeds."

Mrs Warner said that her 25-year-old son Rodney will live there to guard against any possible theft and vandalism.

An electrified fence currently surrounds the sanctuary which also protects the animals from predators like foxes. Her family currently travel everyday from their Enstone home to work on the sanctuary.

She added: "It is not only a security measure but having someone there will provide extra supervision as the animals are giving birth all the time. Chicks are being born every day of the year."

The sanctuary has around 18,000 people visit it every year.

Councillors on the north area planning committee voted in favour of the bungalow, despite it being recommended for refusal by officers.

February 28th 1996

The bungalow for Rodney — Martin Cherry levelled off the site and Maurice and Mark North came to start digging out the footings. It is being approved at each stage by a building Inspector from the District Council. Rodney, Mark and Maurice alternate on a rota, working in pairs. Two loads of concrete and 300mm piping for ditch at the back await. The building regulator inspected the laying of first block work.

April 27th 1996

The building of the bungalow is up to window-sill height and Rod is now doing all the stone cladding while Mark and Maurice build internally. Maurice had taught Rodney how to build in stone. Building regulations have been checked at each stage. It is starting to look good.

May 4th 1996

Building is up to the eaves. All the stone used was hauled by Rodney from the old broken-down barn in our top field.

Maurice and Mark with the footings

Grandson, Micah

Rodney doing internal work

Joinery by Richard Honour

JULY 21ST 1996

I have received a cheque for £35 from the grateful Barley Hill School. *See letter pages.*

JULY 29TH 1996

An eagle owl came to live here. John and Sandra Reid of Edward Street, Banbury, were forced to re-home him because vandals terrorised him. Although our surroundings are idyllic, the owl pined so much for John that one week later I advised taking him home. What a reunion that was — it quite brought tears to my eyes!

I received a letter containing £2 from a young girl visitor living at Bracknell, Berkshire. Bless her!

AUGUST 1996

Sadly Cocky, my Ring-neck Parakeet, died; I have no idea how old he was. This was followed by the death of his beloved cashmere rabbit, Ruby, from a broken heart; the next day Olitwo, my tawny owl, also died of grief. All three had lived in close harmony together with much love for about five years, side by side. As ever, life goes on.

First Emu

SEPTEMBER 18TH 1996

After several phone calls from my vet, Clive Madeiros, a young emu, Graham or Gracie, came to live here. He is about five feet in height with dark feathers, black, shiny eyes and powerful legs, not yet fully grown.

He had been living in a pig-sty in the steep back garden of Grace Cottage, Hook Norton, the home of the other vet, Tony and his wife, Jennifer. He became too big and awesome for their two little boys. Tony and I helped Rodney lift him amid much wriggling and lashing out of legs and wings and managed to manoeuvre him down the slippery steps of the garden. Martin inched his truck up a steep narrow track and I hung on to Rodney's coat as he struggled to contain the bird. Finally, with great relief, the emu was safely in the truck and delivered to the Sanctuary.

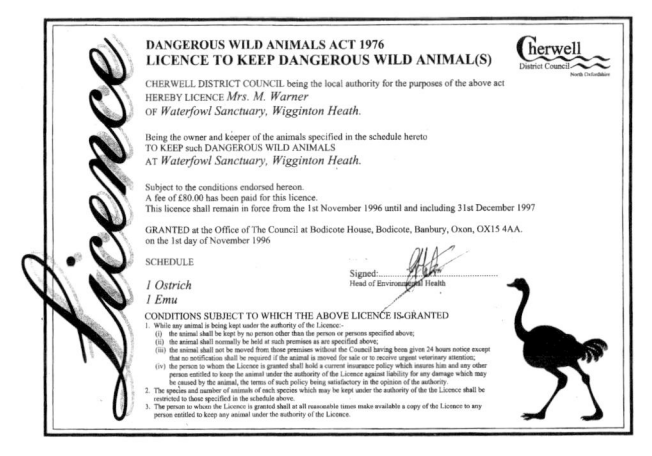

Emu, Graham or Gracie

OCTOBER 6TH 1996

Martin and Rodney erected six-foot chain link fence inside the other paddock eight feet away from the public to enable me to obtain the zoo licence needed to be able to keep emu and ostrich. The vet, Clive Madeiros, sanctions the licence, costing £80, on application to Cherwell District Council, but only when he is completely satisfied that the living conditions of the birds are up to standard and that the birds are fully secure for the protection of the public.

Today I also swopped four Maran hens for two castrated Wiltshire Horn lambs. They look like goats but are much more docile, need no shearing and have lovely shiny coats. They are category 4 rare breeds.

OCTOBER 10TH 1996

Four rheas arrived to join the emu. Martin and John Cherry brought them with their specially formulated grass feed pellets on loan to me as an added interest.

OCTOBER 20TH 1996

Also on loan from vet Clive is a small black South American eight-foot high ostrich. He has promised to bring along a mate for Ozzy soon.

Heidi Gardener from Chesterton is working here five hours a month for one year to attain her *"Care of Animals"* badge for the Duke of Edinburgh's Gold Award.

Helen Gilkes from Chinslade Farm, Cherington, is working enthusiastically and happily part-time. As well as looking after many of the animals, she is now in charge of breeding most of the baby rabbits here. Since July she has kindly made name-plates for most of the animals under her care and is a great help to me.

The designs for rabbit-breeding cages, which Helen produced as part of her project work at Moreton Morell College are shown here.

Since she started, Helen has recorded the rabbit breeding programme in precise detail.

Helen Gilkes, part-time Sanctuary worker

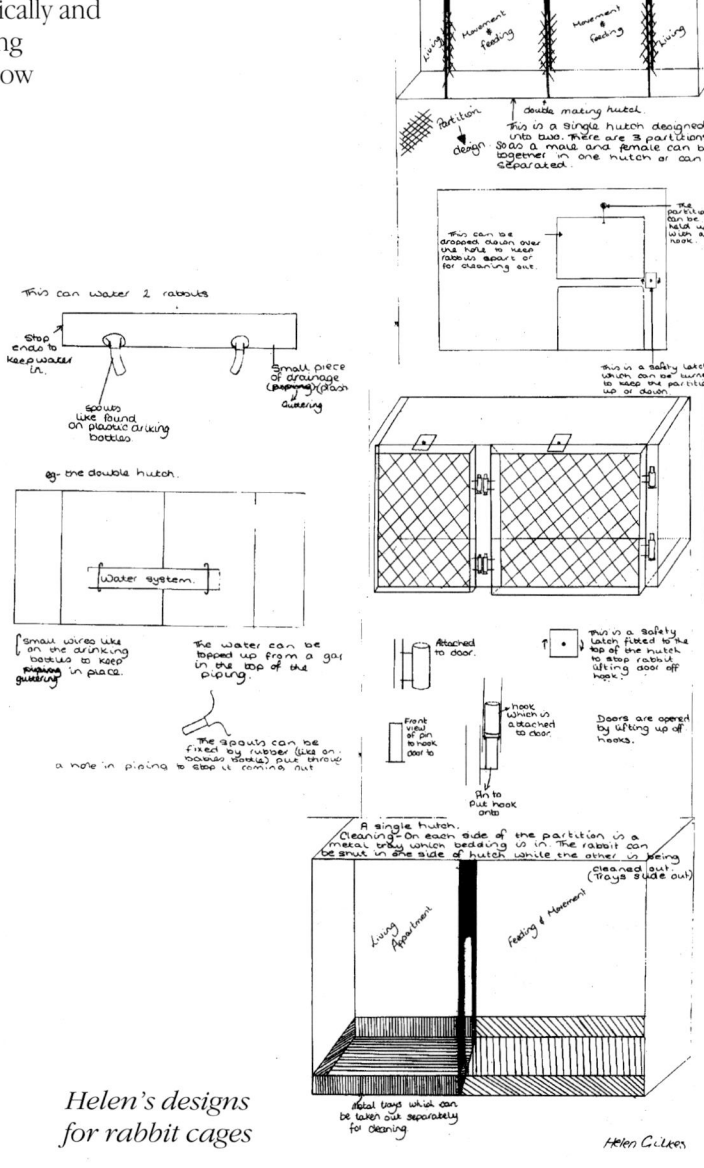

Helen's designs for rabbit cages

Thank you my duck!

NOVEMBER 1996

The *North Cotswold Diamond* gave us some good free publicity in their latest leisure guide. The front cover featured the Hook Norton Brewery Dray, a popular and unusual sight in the area.

The baby barn is proving to be a great attraction, housing piglets, kids, lambs and calves etc.

A trio of pedigree pygmy goats have been purchased from   . They are quite adorable. There are two nannies in kid and one billy.

I have bought some more Rhode Island Red day-old chicks from Aberdeenshire.

A rare Green-legged Rail was brought in recently with a broken leg. Unfortunately it was so badly injured that it died overnight. A local ornithologist said he had seen one on the central pond. Being shy birds they are seldom seen, but there's always a chance you may be able to observe one here at the Sanctuary.

PLACES *to visit*

Ducks, rare breeds in safe environment

■ *Water Fowl Sanctuary and Children's Farm, Wigginton Heath, near Hook Norton, OX15 4LB. Open daily, except Christmas Day, 10.30am until dusk. Telephone 01608 730252.*

Not quite ten years ago Mrs Mabel Warner's love of ducks and waterfowl drove her to plan and open a sanctuary.

She bought some fields in which ponds could be dug, and then to obtain stock, placed an advertisement in the window of her florist's shop in Banbury.

Soon Mrs Warner had a handful of birds to look after and cherish, and thus began an attraction that now delights a regular flow of visitors from around the Globe.

Today there are more than a dozen ornamental duck ponds on a 22-acre site, housing a family of well over 2,000 animals including many rare and endangered species.

Ostriches

The original ducks have been joined by a multitude of other waterfowl, and by rabbits, pigs, chickens, Angora and Pygmy goats, Cotswold and Soay sheep, and by Highland cattle.

Too, there are ostriches, emus, and rheas.

In the ponds are native frogs and newts, sharing the waters with goldfish, koi carp, and terrapins, all adding to a natural and safe environment that has provided study opportunities for schoolchildren and environmentalists alike.

The family grows all the time, and every day visitors may observe new born animals and take pleasure in the gentle handling of chicks, ducklings and baby rabbits.

On site, too, is an interesting and varied flora with meadow land and rich hedgerows in which wild rabbits are much in evidence, and above which kestrels can often be seen hunting.

The sanctuary, which also serves as a refuge for injured and neglected animals of many kinds, has already won an award for conservation.

And Mrs Warner has now obtained a zoo licence enabling her to broaden the scope of her dedicated work.

■ *The sanctuary and children's farm can be found approximately halfway between Banbury and Chipping Norton .... follow the ducks from the A361 or B4035. Free parking for cars and coaches. Admission: adults £3, senior citizens £2.90, children £2. Group discounts are available. Wheelchairs welcome. Practical footwear advisable in wet weather.*

The Boat that Won't Float

FEBRUARY 1997

I have just read in the *Banbury Guardian* that the Banbury Cross Players need a home for their boat that doesn't float. I phoned immediately and they were delighted to give it to the Sanctuary for a small donation.

Unsailable boat needs a home

ARE you after a boat that doesn't float?

If so, the Banbury Cross Players amateur dramatic group might be able to help you out.

The group has just finished work building an ambitious cabin cruiser prop for its stage production of Alan Ayckbourn's Way Upstream.

Jayne Buzzard, of Banbury Cross Players, said: "It's taken weeks to build and we all think it looks pretty spectacular.

"We will dismantle it after our final performance so anyone who wants it as a boat is going to have to put it back together again.

"Alternatively, someone might want to get hold of all the wood we used to make it. The only thing we ask is that people must be able to come and collect it from the theatre."

Anyone who would like to find out more should contact Simon Attwood at the *Banbury Guardian* news desk on (01295) 267310 before the end of the week.

Banbury Cross Players' production of Way Upstream is on now at The Mill Theatre in Banbury until Saturday (February 15). Tickets are available on (01295) 279002.

Sanctuary gives boat a new home

THE boat that doesn't float has found a new home.

The home-made wooden boat was due to be cut up and thrown away last week after the Banbury Cross Players amateur dramatics group finished using it as an elaborate prop in their latest play.

But the owner of Wigginton Wildfowl Sanctuary stepped in at the last minute after reading about the boat's plight in last week's *Banbury Guardian*.

Mabel Warner said: "As soon as I read the story, I thought, 'I know what I can do with that'. It would have been such a shame if someone had just put a chainsaw to it and thrown all that hard work away."

Mabel intends to strengthen and weatherproof the wooden structure and use it as the focal point for the sanctuary's children's play area.

"It'll look fantastic when it's finished and I'm sure the kids are going to love playing on it."

Mabel said she hoped the boat would be ready for the play area in a couple of weeks.

Thank you my duck!

HELLO MY DUCK

By FIONA TARRANT

MABEL Warner has a row of tiny chairs, each with a square of blanket and a notice which says: 'Please sit down'.

It is here that 60-year-old Mabel witnesses sights that make her smile. Tiny tots can sit on the chairs while their parents hand them baby animals to cuddle.

These animals are Mabel's pets and they live at her very own sanctuary. She sold her house ten years ago to buy a shed and the 22 acres around it to become a home for 2,000-plus birds and small animals.

"It was the gamble of my life – a real step into the unknown. I suppose you could say I was foolish or crazy in the extreme, but I love every day of my life now," said Mabel.

Her sanctuary – the Waterfowl Sanctuary and Children's Farm at Wigginton Heath, between Bloxham and Hook Norton, is a favourite with children who want to see animals at close quarters.

"We'd lost touch. I married, but he never did," said Mabel.

Eleven years ago, Mabel and Dennis tied the knot and it's proved to be the perfect union.

Dennis farms at Enstone and grows the corn that Mabel needs to feed her ever-growing menagerie.

"I set the sanctuary up because I love animals and I want them to be happy. A lot of them are old, but here they can see out their days in peace," said Mabel.

Within a minute of arriving at the sanctuary, Mabel had introduced me to a chick born with no feathers ("maybe someone could knit it a little coat," she mused) and a big, fluffy white cashmere rabbit called Snowy.

"Snowy's owners moved away. He's a family pet and he's obviously missing them – we need to find him a good home," said Mabel.

As we walked from one building to another (things have moved on a lot since the sanctuary opened to the public in 1989), we saw peacocks, pigs, Shetland ponies, goats, chickens, ducks and rabbits.

It's not the kind of concrete jungle children see at most zoos. In fact the sanctuary is quite hotchpotch, but Mabel likes it that way.

"It's quite Heath Robinson but the animals all have a lot of room," she said.

Mabel has photographs of her animals taken over the years and she has written her heartwarming story down for a book she hopes to have published called Thank You My Duck.

HOP IT: Duck meets rabbit

"They might hold a gosling, a chick, a rabbit or a kitten – there are baby animals all year round," said Mabel.

"The most important thing is that they learn to be kind to animals. I always say to them 'that little animal loves you' and it melts them," she said.

Setting up the sanctuary has been a long, hard struggle for Mabel, but she has no regrets.

Born in Charlton-on-Otmoor, Mabel was one of seven children and has always loved animals. She and her husband, William Robbins, had florists shops in Botley, Lechlade and Banbury before William died in 1983.

Mabel carried on with the business, but knew she wanted to do something different with her life.

It was then, as a widow, that she met up again with the man who had been her boyfriend when she was just 17. His name was Dennis Warner.

LUCKY DUCKS: Mabel looks upon the sanctuary as a labour of love

Television News

JUNE 1997

Central Television came to film at the Sanctuary following the article by Fiona Tarrant in the *Oxford Mail*. One cameraman and a reporter thoroughly enjoyed their assignment. They stayed for two hours and were reluctant to leave. Their report was shown on *Central News South* on the evening of June 16th 1997.

Visitors look on during the filming

Grandson Lee

Granddaughter Stephanie

My daughter, Wendy, who worked for *Central Television* for several years, gave birth to my fifth grandchild, Charlotte Rebecca, on April 13th 1997. Also, Anne Dawson, one of the *Central News South* Newsreaders, gave birth to a baby girl named Georgina on February 15th, 1997.

WELCOME

TO THE
WATER FOWL SANCTUARY & CHILDREN'S FARM

The Home of Thousands of Happy Animals and Birds

After becoming UNWANTED, UNLOVED, NEGLECTED and ABUSED by mankind every year they gobble up:—

- 25 tonnes of grain
- 600 sacks of seeds and feeds;
- 1000's of bales of hay & straw are used;
- veterinary bills have to be paid.

(to satisfy their varied daily diets)

Your money goes to feed them, care for them and to give them quality of life by building up and improving their living environments.

If you are able to spare a little extra, please do give it to help us in our work.

Thank you,

The Sanctuary could not survive without YOU! Please come again!

NB. Advertising and administration costs are covered by my own personal resources.

This sign welcomes visitors to the Sanctuary

Thank you my duck!

1992-3

Dear Mrs Warner
Neithrop Junior School
Prescott Close
Banbury Oxon

Thankyou for letting us go round the Sanctuary. I enjoyed feeding the animals and holding the baby rabbits.

from Lucy

Dear Mrs Warner Charlbury Primary School
Chadlington
Oxon OX7 3TX

Thankyou for letting us hold the rabbits may 25th and go to wiggington. I really enjoyed it. The best thing was when we went into the adventure play ground. I liked the ducks as well from Emma

---

Charlbury School
26 2 93

Dear mrs warner.
We had a sale and we raised £120 A The stalls were book's raffle toy's lucky dip planet cakes and refreshments. and we hatched some egg's They were Pheasonts egg's 12 of them hatched and I like holding the Rabbit's

best wishes love Nicola

Bloxham primary School
8th July

Dear mrs warner
thank you for letting me feel the baby rabbit. it was lovely and cuddly and the chick crawled off the materials. love from Jessica

---

HORSPATH

WIGGINTON is good fun
I like the white rabbits
Great play ground
Great holy Tortoise
I like ice crem
Naughty goat butted me up the bum
The tower is big
Of we go
Nice Birds

Chadlington Primary School,
Church Road,
Chadlington
Oxon
OX7 3LY

Dear mrs warner
Thankyou for letting us hold the animals
I love it love Rachel

Wooo.

HORSPATH

Wo Love Wiggint.
It is great
Goats do naughty things
Goats eat your clothes
It is big
Nice animals
Terripins are good
On the tower you could see all of wigginton
Naughty goats chewed my paper

by Jamie and Stuart

Thank you my duck! 147

1993-4

Mrs
Dear Warner

8 alan way
george green
slough bucks
SL3 6RA

Thank you for letting as hold the rabbit, mice, chicks and guinea. Thank you for letting as use the Playground too. It was fun looking at the Pig and duck, bird, cow, sheep, etc.

Thank you for a fun day.

from hayley lc

Harriers Ground school,
Bloxham Road
Banbury
20.6.96.

Dear Mrs Warner,
Last Friday our class came Wigginkon. My favourite bit was holding the baby rabbits. The one that I held was a black one. He was nice and fluffy. I saw three peacocks. I saw lots of different kinds of ducks and chicken. I drew a picture of a call duck. I tried to draw a picture of a goose and a picture of a chicken and a duck which I didn't know the name of. I went to the playground it was lots of fun. I bought a feather. I saw two goats and I hid. I hope we will come again. Inside I saw a rabbit with its babies. We also saw an owl and some chicks. Amy picked one up. Terri tried picking one up. Out side again I saw a with peacock and a huge pig. I saw a bird which looking at himself in a mirror. I saw a swan and a white swan. I went all around the place. I saw two shetland ponies and s cows.

from
Alice Birch

Dear mrs Warner
Thankyoo for letting us come to your lovely Water fowl sanctuary. my favourite things were the chipmonks and goats. I Like the goats because they were mischievious. I liked the six week old snowy owl because I have got lots of things. With owls on. I would have liked to take the rabbit I held home I like the pea cock with its fan up.
lots of love from
Jonathan Butler

Harriers ground school
Bloxham Road
Banbury
20.4.94

Dear Mrs Warner, Thankyou for letting us come to your rescue centre. I liked the peacock when it spread its feathers out. I liked the chipmunks. I really liked the rabbits that were running around outside. I liked the chicks and the white peacock. The male goat got a bit angry with us but it didn't charge at us. I liked the black swans. Hammy the hamster was a fidget bottom.

From Tristan

St Patrick's R.C.
(Cashmore Avenue)
Leamington Spa
Warwick Shire
CV31 3EU
29-6-94

Dear Mrs Warner I liked all the animals. My Favourite bird was the owl. The Favourite was the Lovers Seat! Was the big animals a cow or a bull? I Went up a the Tower and I saw Turkeys duck Peacocks Goats one of my Favourite was the Pig, too. My mum wanted a Black Pig but My Dad said no. I like the chicks. It was a hot day. The Terrapins were good I liked the Peacocks Feathers. The Playground was good
yours Sincerely
Thomas Waters

Mollington School

Dear Mrs Warner,
I liked the turkey because he was pretty colours and I could make the turkey gobble

gobble gobble

Love from Becky

Thank you my duck!

1994

Hook Norton C of E School,
Sibford Road,
Hook Norton,
Oxon.
Wednesday 23rd June

Dear Mrs Warner, Thank you for letting me hold the chicks and rabbits I liked it very much.

Love from
Marianne

wrosdy road
Freeland
Oxford OX7 2HX

Dear ms warner
Thk Like you for showing us all the I liked the ducks the best.

From James cook

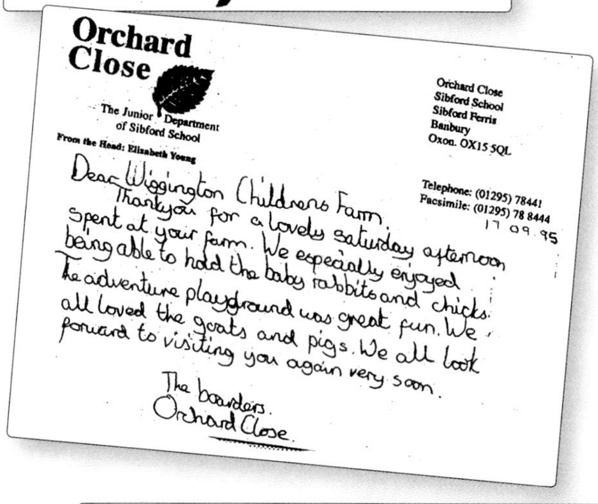

Orchard Close
The Junior Department of Sibford School
From the Head: Elizabeth Young

Orchard Close
Sibford School
Sibford Ferris
Banbury
Oxon. OX15 5QL

Telephone: (01295) 78441
Facsimile: (01295) 78 8444

17.09.95

Dear Wigginton Childrens Farm,
Thankyou for a lovely saturday afternoon spent at your farm. We especially enjoyed being able to hold the baby rabbits and chicks. The adventure playground was great fun. We all loved the goats and pigs. We all look forward to visiting you again very soon.

The boarders.
Orchard Close.

Bletchingdon Parochial (C) Primary School
Weston Road
Bletchington
Kidlington
Oxon OX5 3DH

5/4/95

Water Fowl Sanctuary

Dear Mrs Warner,
Thank you for letting us come, your farm we really enjoyed it. A lot of fun. I would have like us hold and cuddle the animals. I liked the pigs and the mice best of all. had a really good time.

From Adrian

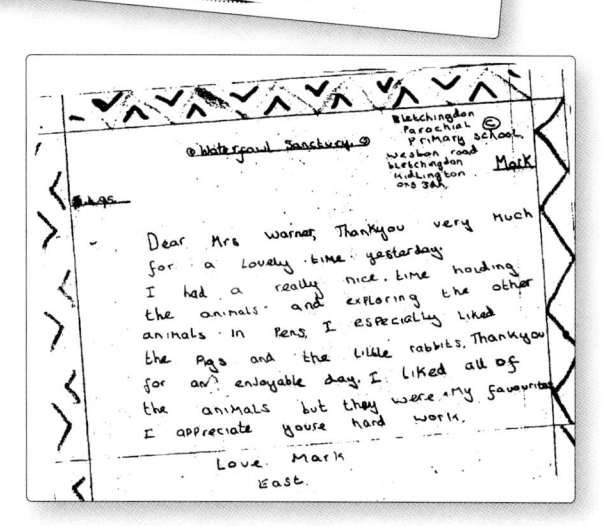

Bletchingdon Parochial Primary School
Weston Road
Bletchington
Kidlington
OX5 3DH

Waterfoul Sanctuary

Mark

Dear Mrs Warner, Thankyou very much for a lovely time yesterday. I had a really nice time houlding the animals and exploring the other animals in pens. I especially liked the pigs and the little rabbits. Thankyou for an enjoyable day. I liked all of the animals but they were my favourite. I appreciate youre hard work.

Love Mark East.

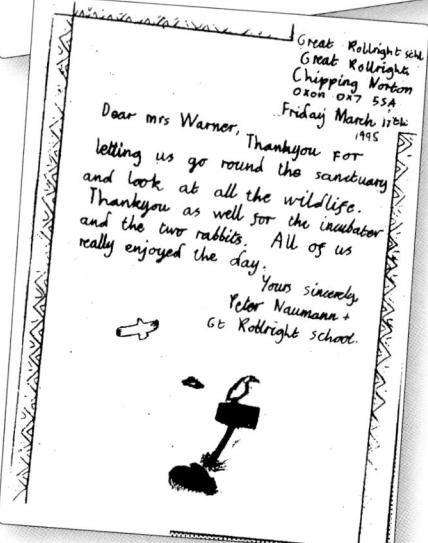

Great Rollright sch.
Great Rollright
Chipping Norton
Oxon OX7 5SA
Friday March 17th 1995

Dear Mrs Warner,
Thankyou for letting us go round the sanctuary and look at all the wildlife. Thankyou as well for the incubator and the two rabbits. All of us really enjoyed the day.

Yours sincerely,
Peter Naumann +
Gt Rollright school.

Thank you my duck! 149

1994-5

Card 1 (Stephen, Sonum, Sarah - 6 years):

Dear Mrs Warner
Thakyou for the eggs I hba a nsa tMie I fka tqt you aenm taf nsa
love from Sa van

duck
hQS
rabbit

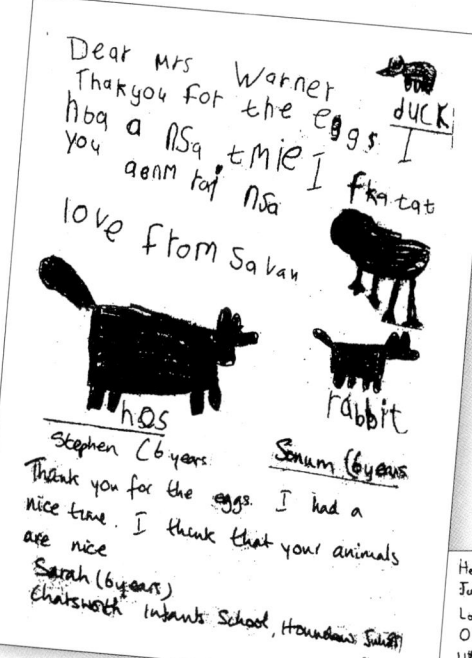

Stephen (6 years)   Sonum (6 years)

Thank you for the eggs. I had a nice time. I think that your animals are nice.

Sarah (6 years)
Chatsworth Infants School, Hounslow

Card 2 (Hannah):

Dear Mrs Warner,
Thank you very much for the lovely trip. We all liked it a lot. I liked the tiny little rabbits best. They are very small and felt like velvet. I felt like their mother when I was holding them because they were so soft.
Love Hannah

Card 3 (Laura Jane):

Headington Junior School
London Road
Oxford
11th June 1995

Dear Mrs Warner
Thank you very much for letting us come and visit The Water Fowl Sanctuary. I liked holding the little duckling with the twisted ankle, it felt nice and soft and did'nt wriggle so much. I also liked holding the little white rabbit which was only six days old and felt very silky.
I hope to take my Mummy and Daddy and younger sister there soon.

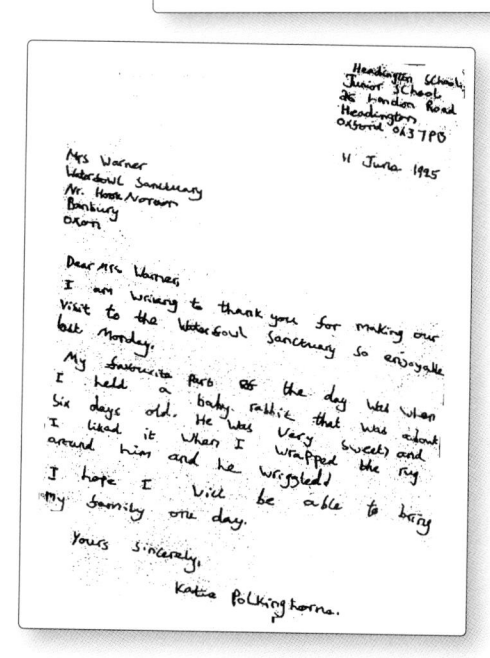

Love from Laura Jane.

Card 4 (Zohreen):

Mrs Warner
I enjoyed the Water Fowl Santuary. The best animal was the Shetland Ponys because they were sweet. Thank you For letting us come to the Water Fowl Santuary.

Love from
Zohreen

Card 5 (Katie Polkinghorne):

Headington Junior School
26 London Road
Headington
Oxford OX3 7PD
11 June 1995

Mrs Warner
Waterfowl Sanctuary
Nr. Hook Norton
Banbury
Oxon

Dear Mrs Warner,
I am writing to thank you for making our visit to the Waterfowl Sanctuary so enjoyable last Monday.
My favourite part of the day was when I held a baby rabbit that was about six days old. He was very sweet and I liked it when I wrapped the rug around him and he wriggled.
I hope I will be able to bring my family one day.

Yours sincerely,
Katie Polkinghorne.

Card 6 (Chloe Cooke):

Headington school
Junior school
form 1s

Dear Mrs Warner,
Thank you for letting us hold the animals.

I liked the very small rabbit because it was very soft and warm and liked it's ear's.

I liked the mummy rabbit too because of its fur.

Love from
Chloe Cooke
xxx

1996

Letter 1:

Barley Hill School
Linden Grove,
Thame, Oxon
21.7.96

Dear Mabel,
Thank you for the wonderful day. When we got back the children made biscuits in the shape of ducks and we sold them at break time.
Enclosed is a cheque for £135-00 to support the work that you do.
Thank you again.
From
Anne Walkinshaw
and the staff and children of Barley Hill School.

Letter 2:

56 Maple Ave
Kidlington
Oxon
OX5 1HN

My name is Kate Adderley.
Here is £12-65 that I have raised for your animals.
I did a sponsored bike ride around my field. 10 yrs old

Kate

Letter 3:

West Witney Primary School
Edington Road
OX8 5FS
June 12th 1996

Dear Mrs Warner,
I enjoyed visiting your farm very much. My best animal was the mice.
Love Charlotte xxxxxxxxxxx
PS I liked the rabbits ***
especially the blak and wight one.

Letter 4:

Rabbits, Guinea pigs, chicks, cockatoo, ponies, pigs, kittens, sheep, peacocks, geese, mice and more,
They are all there.
It's open all year round,
Apart from Christmas day,
When all the animals are having a rest,
In their nice cosy nest.
Mrs Warner looks after them day after day,
Making sure they have plenty of hay,
You can give them all a cuddle,
But walking around, beware of a puddle,
They will love to see you.
You don't have to queue.
And when you go,
Don't cry,
Come again,
We would love to see you,
And we hope you do too.

By Sally Mullins

Letter 5:

West Witney
Primary School
Edington Road
Witney OX8 5FZ
June 12th 1996

Dear Mrs Warner
I liked the chiks
and I liked the rabits
and I liked the pickoks
and I liked the pigs
and I liked the birds
and I liked the duks
and I liked the shons
and I liked the mise
and I liked the ponneys
and I liked the cowes
and I liked the gouts. Love from Jamie

Thank you my duck! 151

1996-7

Our Lady's School
Cowley
Oxford
10th July 1997

Dear Mrs Warner, Thank you for letting me hold the animals. It was very nice of you. I like holding the rabbits and the small mouse and I like holding the kitten. and I like holding the chick and I like holding the ducklin. I like the duckling when he sat on my lap I would like to come agen.

Love from Chenner

Valley Road School
Valley Road
Henley-on-Thames
Oxon RG9 1RR
9/7/97

Dear Mrs Warner

I really enjoyed the farm. I loved the animals. My favourites were the peacocks, kittens, hens, mice, birds and rabbits. I think the peacocks were lovely, the kittens were sweet, the hens were funny, the mice were sweet, the birds were sweet and so were the rabbits. I really really want a pet mouse but my mum doesn't like mice so I might get a kitten or a rabbit.

Thank you for having us. Yours sincerely,
Jessica Cattlin x

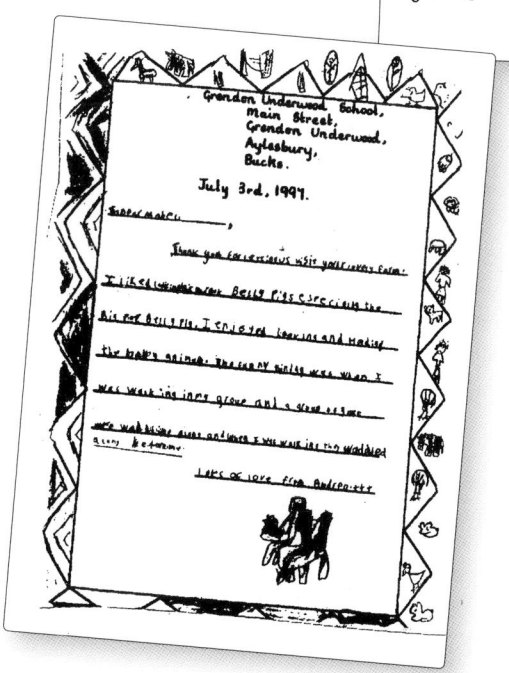

St Johns RC primary School
Ascot Way
Banbury
Oxfordshire
4th July

Dear Mabel,

Thank you for letting St John's Year 3 class come to the Water Fowl centre. I don't know about the rest of my class but I really enjoyed it. I think it was really great the way that you could cuddle the animals but my best animal was the rabbit because I have a rabbit and I think they are really cute. I liked the shop as well. There were new gifts and I really enjoyed seeing all the ducks. I had a great day out.

Yours sincerely
Jacob Payne

Grendon Underwood School,
Main Street,
Grendon Underwood,
Aylesbury,
Bucks.

July 3rd, 1997.

Dear Mabel,

Thank you for letting us visit your lovely farm.
I liked watching the belly pigs especially the big fat belly pig. I enjoyed looking and holding the baby animals. The funny thing was when I was walking into the group and a head of me was waddling ever so low and I was walking waddling along behind me.

Lots of love from Andrew xxx

Thank You Thank You
A VERY SPECIAL THANK YOU

Hill View School
Banbury
10.9.96

Dear Mrs Warner

Thank you verry much for the Rabbit. We have called him Roger. He is cuddley and sweet. Mrs Hewitt chooses two people to stay in and hold him. He hops around the classroom.

love
Kelly Marie Hawley

Thank You Thank You

Mabel at the reception desk

Picnic by the Shepherd's Hut

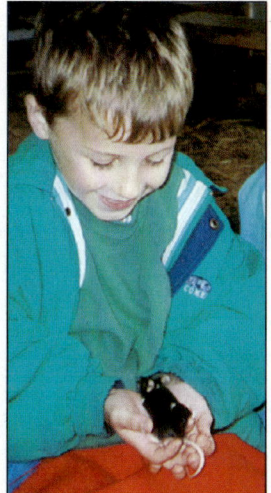

Above from left to right:
Vicky Nunn, Racheal East, Sam Dix, Holly Low and Casima Meakin, Finmere School

Left:
My special friend, Sally Mullins, Horspath

Right:
Samantha and Ricky Dee and Alex Brown from Brackley, Northants, 1997

Left:
Hayley and friends from Yarnton School

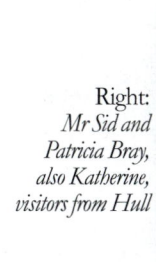

Right:
Mr Sid and Patricia Bray, also Katherine, visitors from Hull

I

Left:
My old friend Aunty Loe from Lechlade and Janet

Right:
I did enjoy cuddling the little lamb

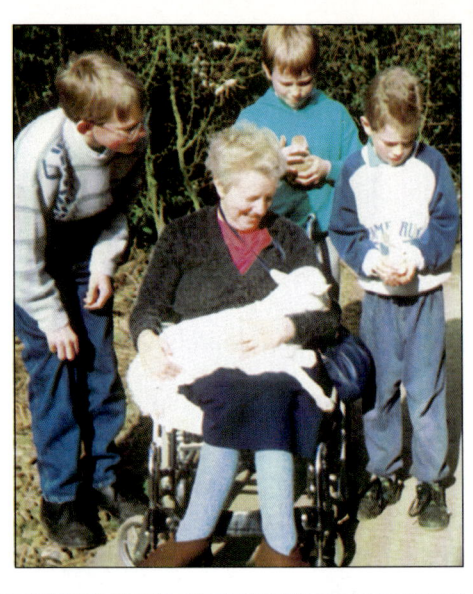

Left:
Visitors enjoying a picnic

Right:
My friends Joyce and John's grandchildren James and Lucy feeding lambs

Left:
I'll take this one, my duck, the Boswell children from Bloxham

Right:
Samantha Walman

J

Jill from West Bar Vets and myself Tudor Photography

Jonathan from Hook Norton, who first introduced me to Australian Fruit Bat Mice

Faithful supporters

Harry with Jimmy, our Lesser Crested Cockatoo, with the ear-piercing screech

My grandson, Micah, with Jimmy

Woodfolk 1990

Tony Baldry, M.P.
"You won't bite me, will you?"

Rodney

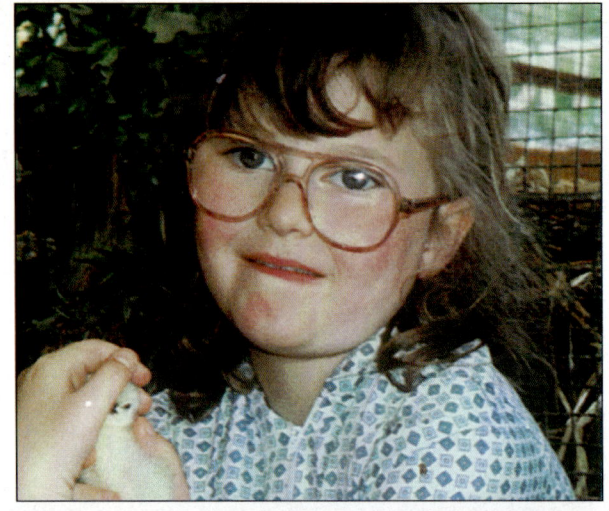

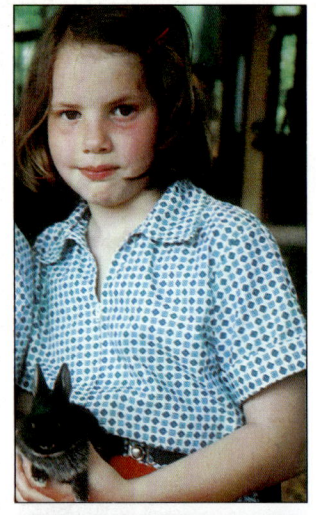

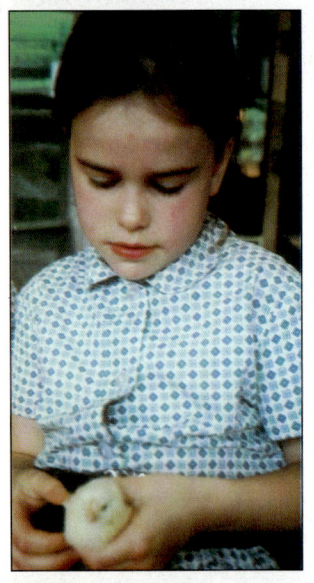

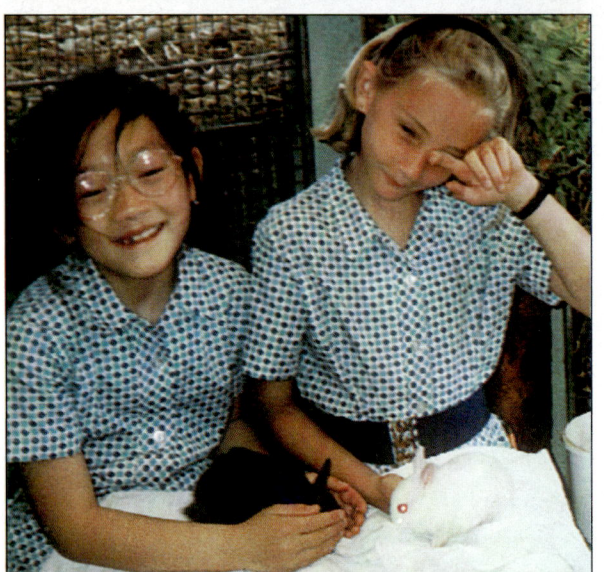

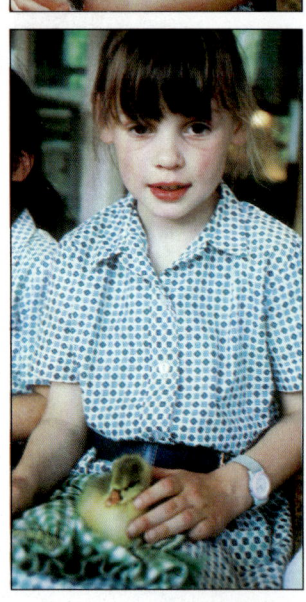

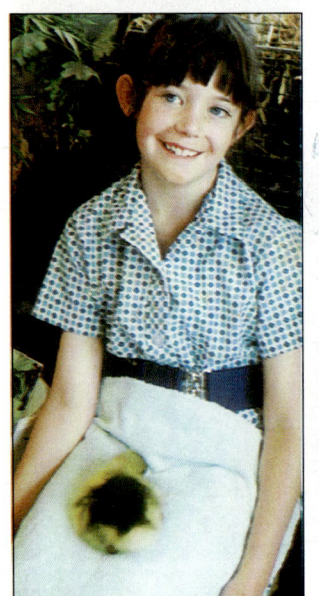

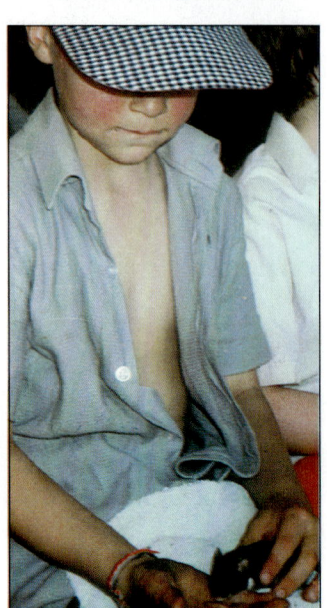

Children from Headington School
Photos by Sheila Slater

Left:
My friends from Italy visit every year without fail, Vanessa and Beatrice Marcini, Mum and Gran

Right:
Happy visitors

Left:
A large, beautiful family of animal lovers

Right:
Visitors feeding spring lambs

Above:
Doug from Yorkshire

Left:
Happy schoolchildren

Right:
Tessa Terry with her son Harry Bowerman

*Francesca in wheelchair, a lovely little girl with a
beautiful smile, suffering from a severe illness —
with her friend Jude*

Here at Wigginton baby rabbits are called "bunnies" and
baby cats are called "kittens". How can they possibly
both be called "kittens" as some people believe?

Coombe Schoolchildren.
By kind permission of Press House Publications, Hook Norton

Highland Cattle and Soay Sheep

Penny and Eddy in quiet conversation Rod Smallman

Sara the Jacob Sheep with Donk the Balvan Sheep

Winniepiglet, our Oxford Sandy and Black Pig

Poppy, the Golden Guernsey Goat

Aaaaah …!

Egyptian Goose Photo: Rod Smallman

Sebright Cock Bantam

European Shellducks Photo: Rod Smallman

Indian Blue Peacock Photo: Rod Smallman

Ozzie and Graham: Ostrich and Emu

Entrance to the Sanctuary

Thank you my duck! 153

A sample of other letters

OXFORDSHIRE COUNTY COUNCIL

SS Philip and James's First School
Leckford Road
Oxford OX2 6HX

2nd April 1993

Dear Mrs Warner,

We have been very pleased with the Brinsea incubator we borrowed from you.* We had 12 chicks from 24 eggs. The children also loved their visit to Wigginton Waterfowl Sanctuary.

We have all learnt much from this project which I am sure has made a very positive, lasting impression on the children.

Yours sincerely,

Ruth Jens.

* This was easy, compact and a clean incubator to use, and very suitable for classroom use. The self-turning device was invaluable.

OXFORDSHIRE COUNTY COUNCIL

GREENMERE COUNTY PRIMARY SCHOOL
MEREHEAD ROAD
DIDCOT
OXON OX11 8BA

March 26th 1993

Dear Mrs Warner

This is just to say thank you very much for our visit last Tuesday. Our day was most enjoyable and the children welcomed the experience of handling all your "babies". We hope the Sanctuary continues to prosper and expand. Our thanks are again for all your help and kindness.

Yours sincerely,
Cathy Shank

EMMANUEL CHRISTIAN SCHOOL OXFORD

Tel: 0865 311828

Middle Way,
Summertown,
Oxford,
OX2 7LQ

13 July 1993

The Warden,
Water Fowl Sanctuary,
Wigginton Heath,
Nr. Hook Norton,
Banbury,
Oxon OX15 4LB

Dear Warden,

Thank you so much for giving us such a good day yesterday. The children enjoyed their visit to your sanctuary immensely and were fully occupied and fascinated for the best part of the whole day.

May I say what a good resource your centre is for our school. As a christian school we are very interested in environmental issues and found your worksheets based on National Curriculum attainment targets excellent. It is very good for our children to develop observation and data handling skills in "the field" as well as the laboratory or classroom and I will be recommending to other members of staff that they use your centre for other Keystages.

The children especially enjoyed the close contact with animals and we teachers much appreciated the friendly "child-proof" set up.

Thanks again.

Yours sincerely,

Sally Stokes
Class Teacher

Thank you yet again for a very successful trip to Wigginton. The children learn so much about animal care and the most important part, handling young animals. They always came away having made at least one new friend and for children of 8 and 9 that is very important.

We used the worksheets for the first time and they were excellent. They gave the children a focus and a starting point for important discussion about habitats, identification, differences etc. I could also choose different worksheets according to the different ages that I have.

I shall be using Wigginton as a curriculum resource for sure more in the future and I know the children will learn far more from being practically involved than a shelf full of books.

Best wishes Sheila Franklin
Teacher Lower Juniors (Horspath Primary)

Letter 1 (handwritten)

Tysoe School
Weds 12th May

Dear Mrs Warner,

I would like to thank you for welcoming us all to the Waterfowl Sanctuary last Wednesday. The children all enjoyed themselves and got a lot out of the visit. I enclose some of their letters for you to read.

Yours Sincerely
Alison K. Sayer

Letter 2 (Southern Tourist Board)

40 Chamberlayne Road, Eastleigh,
Hampshire SO5 5JH Telephone:
National - Eastleigh (0703) 620006
International - + 44 703 620006
Fax: 0703 620010

29th April 1993

Mrs M Warner
Water Fowl Sanctuary
Rectory Farm Bungalow
Church Enstone
Oxford
OX7 4NN

Southern Tourist Board

President: The Lord Montagu of Beaulieu
Vice-President: Councillor F.A.J. Emery-Wallis
Chairman: Michael W. Green
Vice-Chairman: Councillor Dudley Keep

Dear Mabel

SOUTHERN TOURIST BOARD
STAFF PRODUCT EDUCATIONAL VISIT 23/24 APRIL 1993

On behalf of all staff at the Board, I am writing to thank you for making our recent visit both informative and enjoyable.

Although we could not spend as much time as we would have liked at any venue, it certainly gave us a taste of a most beautiful corner of England which Southern Tourist Board is proud to represent.

Many thanks again for your cooperation, we look forward to meeting you again in the not too distant future.

Yours sincerely

Sindie Hughes

DEBBIE HORLOCK
Information Manager

Letter 3 (handwritten testimonial)

Testimonial in Support of The Water Fowl Sanctuary, Wigginton Heath. 30th March 1993

I would like to recommend highly this Sanctuary as being of excellent educational value to school children of varying ages. My class of 7 and 8 yr. olds were absolutely fascinated to handle, observe and feed the many varied large and small animals. We used a number of the prepared worksheets (which we found of a very high standard and interestingly varied) to cover Science A.T. 2, Strands 1, 2, 3, 4 – Life & Living Processes. We were also able to cover Science A.T.1, Strands 1 + 2 plus some of the Maths A.T.'s (i.e.) Data Handling & Number.

Mrs. Warner runs the Sanctuary in a very friendly way, tells the children about the animals, answers their questions & has everything prepared for visits with clean washing + toilet facilities, little seats with cloths and finally something to take home.

I knew the children would enjoy & learn a great deal from their visit but even so I didn't envisage the tremendous enthusiasm shown by all the children even those whose concentration in the classroom may be limited. Every child remained ascinated for the whole of our day trip, although it rained almost non-stop! And they were still asking to hold just one more baby rabbit or chick when it was time to go back to school.

This is the only place I know of which can offer all these things plus a Nature/Study Field, Hedgerow & Ponds in a beautiful rural setting. We will plan to visit the Sanctuary every year. The children all said it was the best trip they'd ever been on.

Cutteslowe First School
Wren Road
Oxford

Julie Eaton
Class Teacher

Letter 4 (St Ebbe's School)

OXFORDSHIRE COUNTY COUNCIL
St Ebbe's C.E. (Aided) First School

Whitehouse Road
Oxford OX1 4NA
Tel: 0865 248863
Headteacher:
Sue Matthew

May 26th 1993

Dear Mrs. Warner,

We are writing to say how much we enjoyed our visit to your Water Fowl Sanctuary yesterday.

You gave us such a friendly welcome and the children's worksheets and map were most useful and appropriate. We were particularly pleased that the children could see the animals and birds at such close proximity and in complete safety could watch them using the ponds. The observation tower was ideal in helping them to see the plan of the sanctuary and supported our previous work, on early map making and plans.

The cuddling time was very special for each child and was another experience which they have not had on any of our other visits to wildlife parks. Today the children are still talking excitedly about the experience and as a result of their child's reaction to the visit many of the parents are wanting to take their children to the Wigginton Water Fowl Sanctuary again.

Thank you for your care, patience and expertise in handling the children as well as the animals in the endless cuddling sessions! The children learned a lot from their first hand experience and we are already looking forward to our next visit.

Yours sincerely,

(Mrs) Dorothy Field (Deputy Headteacher)

Thank you my duck!

Mollington School,
Mollington
Banbury,
Oxon.
19th March 1991

Dear Mrs Warner,

We all enjoyed our visit so much on Tuesday that the children wanted to write a letter to thank you. They all had a special memory about the visit and a favourite animal to draw. Thankyou for being so patient and sharing your love of animals and birds with all of us.

Kind regards,
Mrs Linda England.

HEADINGTON SCHOOL • OXFORD OX3 7PB
JUNIOR SCHOOL
Telephone: OXFORD (0865) 61774

12.6.95

Dear Mrs Warner,

As you will see from the enclosed letters from my class, we once again enjoyed our visit to The Water Fowl Sanctuary.

It is such a delight to visit such a site, in beautiful countryside, with everything arranged with the best interests of both animals and children at heart.

Our children enjoyed it so much that they all want to come back and bring their parents.

With all good wishes for the future. Yours sincerely,
Marta Sykes.

HEADINGTON SCHOOL, OXFORD LTD.
Reg. Office: Ellenborough House, Wellington Street, Cheltenham
Registered in England No. 141076

Greenway First School
Crossways,
Berkhamsted.
Herts.
16th June '95

Dear Mrs Warner,

We would like to thank you for giving the Greenway First School children the opportunity to visit your sanctuary & particularly for allowing such a valuable "hands-on" experience.

The children really enjoyed their visit & were able to see Nature at work.

The video was particularly helpful in showing the children what they could expect from their visit. They all said how much they would like to come again & have asked their mums & dads to bring them.

We particularly enjoyed the relaxed, caring atmosphere & the natural surroundings.

The worksheets offered are well related to the National Curriculum & the topics covered a wide range within Key Stage 1.

We will recommend the sanctuary to the schools in our area & publicise your good work as we feel it is very worthwhile.

Thank you for all your efforts once more.

Yours
Lydia Clark
Karen Denholm

(Class Teachers)

> Community of St Clare St Mary's Convent Freeland Witney Oxon OX8 8AJ
>
> Aug. 13th 1997
>
> Dear Mabel,
>
> I hope to be able to visit with my friends at the end of the month.
>
> Meanwhile, an update on Goose Ling and her husband Dong who came from the Sanctuary in the autumn of 1995. Ling started sitting on her eggs on June 26th and on July 29th this little one had hatched. I call the gosling 'Wiggin'. For a couple of days & we kept Wiggin in the house, but it was rather a headache keeping an eye on gosling & cats, so on Aug. 2nd we put Wiggin back with Ling & Dong and buried the other dozen eggs which were giving no signs of life.
>
> Wiggin grows fast, all day is spent eating grass & lettuce!
>
> with all good wishes & prayers
> S. Damian

Sr. Damian's goose, Ling, and gosling, Wiggin

Tara's gentle touch helps a baby bird

TARA the Golden Labrador came to the rescue of a three-day-old blue tit chick which had fallen from its nest.

The gentle hound discovered the chick lying on the ground in High Street, South Newington, last Thursday and waited with it until her owner, Jean Tarver, arrived.

Jean, who lives in the village, wrapped the bird in a tissue and took it to Bloxham, but found the vet's surgery closed.

"I was going to put Tara in the car but she was sniffing round something, and when I looked closer I could see it was a baby bird," said Jean.

"I thought 'I can't leave it' so I picked it up. When I found the vet was closed I had to take it to the doctor's with me."

Following the receptionist's advice she took it to the Waterfowl Sanctuary at Wigginton Heath, where the tiny bird was put into an incubator.

Sadly the chick later died of shock.

How the Chipmunk got its Stripes

This species of chipmunk originates in Siberia where a story is told about how the chipmunk came to have its beautiful stripes —

A large bear once woke up from it's winter sleep, and as he had not eaten for so long he was extremely hungry and it set off searching for food.

He searched, but couldn't find any anywhere. Finally he came to an old tree stump, and wondered whether there might be some honey inside it. He began to dig.

Suddenly a chipmunk popped out. and asked what he was doing, saying that the tree was his home.

"I'm sorry" said the bear "but I am very hungry and I thought I might find food in here."

The chipmunk muttered and dashed away thinking: "What kind of animal is it that doesn't prepare food ready for itself when it wakes up in the Spring?"

A little later the bear was very surprised when the chipmunk returned carrying food from his own winter supplies. These he offered to the bear, who was so delighted that it reached out and stroked the little chipmunk. His claws left dark stripes down the chipmunk's back which are still there today to show the world what a generous creature it is.

Female Rabbit Behaviour

There are many excellent books on the care of pet rabbits. Unfortunately we haven't come across one that covers this particular aspect of rabbit care, so we offer the following advice from our own experience.

At approximately 7-9 months of age, (or even earlier), some rabbits change their character, indicating their need to reproduce. If this affects your rabbit, she will begin to pluck fur from her underside and sides to line a nest. She will also carry straw and hay to a corner of the hutch, and stamp her feet.

If the owners disregard these signs and do not arrange for her to be mated, she may begin to attack her feed bowl when it is put in the hutch. She may resort, in desperation, to nipping sleeves or even hands. She will become increasingly unsociable and may no longer enjoy being cuddled and stroked.

Our advice is to allow her to have one litter with a carefully chosen, good-natured buck. We believe that all should then return to normal.

Always introduce the doe into the buck's enclosure, not the buck to the doe.

Did you know that?......

1. I am led to believe that thistles harbour fleas which carry myxomatosis, the killer disease that attacks rabbits usually around harvest time each year.

2. To help a chick out of it's shell brush cod-liver-oil on to the egg over the pipping hole. Young chicks can be fed chopped hard boiled egg and chopped dandelion leaves.

3. Ducklings hatched in incubators can drown in deep water. Naturally, the mother duck spreads a dressing of oil from an oil gland beneath her body all over her young, which makes them waterproof. Also static electricity is missing by not being transmitted from the mother, so some folk stroke the duckling with a silk material to rectify this, and they also feed them cod-liver-oil.

   Rain can kill ducklings with no oil on their feathers. You've probably heard the saying: *"A dying duck in a thunder storm"*. This is quite true.

4. From Spring 1995 you no longer have to test for salmonella in eggs for retail sale.

5. At night foxes attack peahens more than peacocks because peahens roost lower than peacocks and can therefore be reached more easily.

6. Peacocks drop their tail feathers during June, July and August after the mating season, and grow new ones the following spring ready to attract a mate.

7. The fluffy, untidy looking rare breed Sebastopol Goose is the original Pantomime Goose. I am led to believe pantomimes began in Russia where this quiet, docile goose was the perfect star in 'Mother Goose' productions.

8. Folk born in the lower villages around Otmoor in Oxfordshire were known as Otmoor Goslings, (including me), and were said to have webbed feet. *(An old wives tale)*.

9. Female rabbits when needing to reproduce sometimes change their character and become quite unfriendly. The best solution is to find a buck of the same size and allow her to have one litter at least.

10. Cashmere fibre is 'fur' from Cashmere rabbits and 'hair' from Cashmere goats.

11. Angora fibre fur/wool comes from Angora rabbits and Angora goats.

12. Angora goats come from Africa, New Zealand, and Canada and have long curly, silky soft fleeces, almost touching the ground before they are shorn each year.

13. There is a species of duck called Buff Orpington, and also a species of chicken called Buff Orpington.

14. A "dillon" is the name for the smallest pig of the litter in Oxfordshire; in Wiltshire it is called a "winnick".

15. The term of endearment "my duck" is similar to "hen" or "chuck" around Liverpool and the north-east.

Thank you my duck!

The Water Fowl Sanctuary

Wigginton Heath
Nr Banbury
Oxon.

Rescue Centre

TEL: 0608 730252

Open Daily 10.30 a.m. - Dusk
Except Christmas Day

Come and share in the joy of our
FEATHERED FRIENDS
Wildlife & rare breeds conservationists

admission
Adults £2 Children £1
10% Discount on parties of 10 upwards

People in wheelchairs especially welcome
layout designed with you in mind

How to find us see reverse
Banbury approx. 6 miles, Chipping
Norton approx. 8 miles, Hook Norton
approx. 2 miles

SURPLUS STOCK FOR SALE

1990

Also goats sheep lambs. rabbits, guinea pigs etc.
Protected flora and fauna
Adventure Playground & Nature Trail
Practical clothing & footwear advisable

The Water Fowl Sanctuary

Wigginton Heath
Nr Banbury
Oxon.

Rescue Centre

TEL: 0608 730252

Open Daily 10.30 a.m. - Dusk
Except Christmas Day

Come and share in the joy of our
FEATHERED FRIENDS
Wildlife & rare breeds conservationists

admission
Adults £2 Children £1
10% Discount on parties of 10 upwards

People in wheelchairs especially welcome
layout designed with you in mind

How to find us see reverse
Banbury approx. 6 miles, Chipping
Norton approx. 8 miles, Hook Norton
approx. 2 miles

SURPLUS STOCK FOR SALE

Two thousand birds and animals on view.
Also, goats, sheep, lambs, rabbits, guinea pigs etc.
Protected flora and fauna
Adventure Playground & Nature Trail
Practical clothing & footwear advisable

The Award Winning Water Fowl Sanctuary

Wigginton Heath
Nr. Hook Norton
Banbury, Oxon.

Rescue Centre

TEL: 0608 730252

Open Daily 10.30 a.m. - Dusk
Except Christmas Day

Come and share in the joy of our
FEATHERED AND FURRY FRIENDS
Wildlife & rare breeds conservationists

admission
Adults £2 Children £1
10% Discount on parties of 10 upwards

People in wheelchairs especially welcome
layout designed with you in mind

How to find us see reverse
Banbury approx. 6 miles, Chipping
Norton approx. 8 miles, Hook Norton
approx. 2 miles

YOUNG STOCK FOR SALE

Two thousand birds and animals on view.
goats, sheep, lambs, rabbits, guinea pigs etc.
Protected flora and fauna
Adventure Playground & Nature Trail
Practical clothing & footwear advisable

An Oxfordshire Special Conservation Award Winner 1990.

The Award Winning Water Fowl Sanctuary

Wigginton Heath
Nr. Hook Norton
Banbury, Oxon.

Rescue Centre

TEL: 0608 730252

Open Daily 10.30 a.m. - Dusk
Except Christmas Day

Come and share in the joy of our
FEATHERED AND FURRY FRIENDS
Wildlife & rare breeds conservationists

admission
Adults £2 Children £1
10% Discount on parties of 10 upwards

People in wheelchairs especially welcome
layout designed with you in mind

How to find us see reverse
Banbury approx. 6 miles, Chipping
Norton approx. 8 miles, Hook Norton
approx. 2 miles

YOUNG STOCK FOR SALE

Two thousand birds and animals on view.
goats, sheep, lambs, rabbits, guinea pigs etc.
Protected flora and fauna
Adventure Playground & Nature Trail
Practical clothing & footwear advisable

An Oxfordshire Special Conservation Award Winner 1990.

Thank you my duck!

The Award Winning Water Fowl Sanctuary

Wigginton Heath
Nr. Hook Norton
Banbury, Oxon.

Rescue Centre
TEL: 0608 730252

Open 10.30 a.m. - Dusk
(Earlier by arrangement)
Open **364 days** a year

COME AND CUDDLE AND PET OUR
BABY BUNNIES, CHICKS, DUCKLINGS, LAMBS ETC.
UNDER SUPERVISION

Admission
Adults £2   Children £1
10% Discount on parties of 10 upwards
Free preparatory visits for Schools and Groups
For Education Pack please send an A4 S.A.E.
with two first class stamps

Banbury approx. 6 miles,
Chipping Norton approx. 8 miles,
Hook Norton approx. 2 miles

YOUNG STOCK FOR SALE

Two thousand birds and animals on view.
Goats, sheep, lambs, cattle, pigs and rabbits etc.
Protected flora and fauna
Adventure Playground & Nature Trail
Practical clothing & footwear advisable

An Oxfordshire Special Conservation Award Winner 1990.

WATER FOWL SANCTUARY & CHILDREN'S FARM

Between Bloxham & Hook Norton
Wigginton Heath, Banbury, Oxon.
☎ **01608 730252**

Come and cuddle and pet our baby bunnies, chicks, ducklings etc. under supervision

2000 rare birds and animals to see & enjoy!

WE ARE HERE — Wigginton Heath
TADMARTON — B4035
MILCOMBE
To M40 Junct 11 BANBURY
A361 BLOXHAM
HOOK NORTON
SOUTH NEWINGTON
WIGGINTON
HEMPTON
To CHIPPING NORTON
TOURIST SIGNS on approach roads

Open 364 days a year 10.30 am — 6 pm
1996 Prices: — Adults £3.00   Children £2.00
(2yrs to 15yrs accompanied by paying adult)
Golden Oldies £2.50
Coachload pre-booked £70.00
FREE 15-MIN. VIDEO / PREP. VISITS
FOR TEACHERS & GROUP LEADERS
Practical Footwear advisable

Member Southern Tourist Board
THE HEART OF ENGLAND — TOURIST BOARD —
O.S.C.A.

WATER FOWL SANCTUARY & CHILDREN'S FARM

Wigginton Heath
Nr. Hook Norton, Banbury, Oxon.
☎ **01608 730252**

Open 364 days a year
10.30 am — 6 pm

Come and cuddle and pet our baby bunnies, chicks, ducklings etc. under supervision

Adults £2.50
(O.A.P.'s £2.00)
Children £1.50
Coach Drivers' Concessions

1995
3000 creatures to see & enjoy!

SHAKESPEARE COUNTRY
TADMARTON — B4035
WE ARE HERE — Wigginton Heath
MILCOMBE
To M40 Junct 11 BANBURY
A361 BLOXHAM
HOOK NORTON
SOUTH NEWINGTON
WIGGINTON
HEMPTON
To CHIPPING NORTON
THE COTSWOLDS
PARTY DISCOUNTS

Teachers' Packs available – with Free Visits

WATER FOWL SANCTUARY & CHILDREN'S FARM

Wigginton Heath,
— Between Bloxham & Hook Norton —
Banbury, Oxon. OX15 4LB
☎ **01608 730252**
e-mail: watfowluk@aol.com

Come and cuddle and pet our baby bunnies, chicks, ducklings etc. under supervision

100's of friendly birds and animals waiting to meet you!

WE ARE HERE — Wigginton Heath
TADMARTON — B4035
MILCOMBE
To M40 Junct 11 BANBURY
A361 BLOXHAM
HOOK NORTON
SOUTH NEWINGTON
WIGGINTON
HEMPTON
TO CHIPPING NORTON
TOURIST SIGNS on approach roads

Open 364 days a year 10.30 am — 6 pm or dusk
1998 Prices:- Adults £3.00   Children £2.00   OAPs £2.90
(1 yr to 15yrs accompanied by paying adult)
Schools & Playgroups Half Price
Coaches & Cars Free Parking
FREE 15-MIN. VIDEO / PREP VISITS
FOR TEACHERS & GROUP LEADERS
Practical Footwear advisable

Member Southern Tourist Board
THE HEART OF ENGLAND — TOURIST BOARD —
O.S.C.A.

Thank you my duck! 161

WATER FOWL SANCTUARY & CHILDREN'S FARM

Wigginton Heath, Banbury, Oxon. OX15 4LB
— Between Bloxham and Hook Norton —

☎ **01608 730252**

So Much More

There is ample space for picnics, including a newly created covered area. Seating is conveniently placed for those who simply want to sit and watch the birds for a while.

Situated in the heart of the Oxfordshire countryside, the Sanctuary occupies a gently sloping south-facing site. On this naturally wet hillside, the network of ponds has been developed from two ancient ponds which once served the needs of livestock.

To the right of the large central pond, a line of ponds follows an old, established hedgerow flanking the roadside. To the left of the main pond are the covered flight runs where many unusual specimens are to be found.

A new generation of miniature rabbits is being born here called "Mabel's Midgets". They are bred from the best-natured and tiniest parents, staying quite small when fully grown, making ideal pets for children of all ages and are available all year round at reasonable prices.

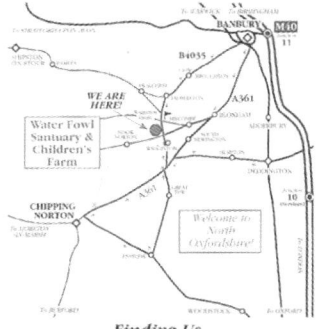

Finding Us

Banbury is about a mile west of M40 junction 11.
The Sanctuary is roughly half-way between Banbury and Chipping Norton—just follow the ducks from the A361 or the B4035.

- 16 ORNAMENTAL DUCK PONDS
- SNACKS AVAILABLE
- FREE PARKING: Cars & Coaches
- WHEELCHAIRS WELCOME

Member Southern Tourist Board

OXFORDSHIRE SPECIAL CONSERVATION AWARDS

THE HEART OF ENGLAND TOURIST BOARD

This Commendation is made to reward an outstanding contribution to Conservation of the Environment

Admissions 1997

ADULTS £3.00, (OAP'S £2.90) CHILDREN £2.00
10% discount for parties of 10 or more (pre-booked).
Early opening by arrangement,
otherwise from 10.30am until 6pm (or dusk).
7 days a week except Christmas Day
Practical footwear advisable in wet weather.

Paragon Print & Design · Milcombe · Banbury · Oxon.

Over 3000 creatures to see and enjoy!

The family-run sanctuary, south-west of Banbury, is well-known for its outstanding contribution to the conservation of rare breeds, wildlife and the surrounding countryside.

Established in 1989, it has become a haven for a great variety of birds and many other interesting and unusual creatures.

WATER FOWL SANCTUARY & CHILDREN'S FARM

Wigginton Heath,
— Between Bloxham and Hook Norton —
Banbury, Oxon. OX15 4LB

☎ **01608 730252**

http://www.visitbritain.com
e-mail: watfowluk@aol.com

Children should always be supervised by an adult when handling the baby animals and birds

So Much More

There is ample space for outdoor picnics in shady glades and there are also several covered areas. Plenty of seating is conveniently placed around the Sanctuary for those who simply want to sit and watch.

Situated in the heart of the Oxfordshire countryside, the Sanctuary occupies a pleasant south-facing site. The network of ponds, which was developed from two ancient pools once serving the needs of livestock, meanders along beside an old, established hedgerow flanking the roadside. To the left of the main pond are the covered flight runs where many unusual poultry specimens are to be found.

A new generation of miniature rabbits is being born here called "Mabel's Midgets". They are bred from the best-natured and tiniest parents, staying quite small when fully grown, making ideal pets for children of all ages and are available all year round at reasonable prices.

Finding Us

Banbury is about a mile west of M40 junction 11.
The Sanctuary is roughly half-way between Banbury and Chipping Norton — follow the brown & white duck signs from the A361 or the B4035.

- 12 ORNAMENTAL DUCK PONDS
- FREE PARKING: Cars & Coaches
- WHEELCHAIRS & GUIDE DOGS WELCOME
- GOOD FOOD AVAILABLE NEARBY

OXFORDSHIRE SPECIAL CONSERVATION AWARDS

Member Southern Tourist Board

THE HEART OF ENGLAND TOURIST BOARD

This Commendation is made to reward an outstanding contribution to Conservation of the Environment

Contact: Mabel Warner

Admissions 1998

Adults £3.00, (OAP's £2.90) Children (1-15yrs). £2.00
Schools and Playgroups Half Price

Early opening by arrangement,
otherwise from 10.30am until 6pm, or dusk
7 days a week except Christmas Day.
Practical footwear advisable in wet weather.

Paragon Print & Design · Milcombe · Banbury · Oxon.

Hundreds of friendly animals and birds waiting to meet you!

COME and CUDDLE and PET our baby bunnies, chicks, ducklings, lambs, guinea pigs, kids, chinchillas etc!

As Seen on Television 1997

*Established in 1989, this family-run sanctuary, has been voted the **best Oxfordshire Family Attraction of the Year** for both 1997 and 1998 in "The Good Guide to Britain".*

"Let's go with the Children!"

For four years this advertisement, in full colour, replaced the usual back-page one for Windsor Castle following the fire and subsequent restoration of the castle.

THE GOOD GUIDE TO BRITAIN 1997

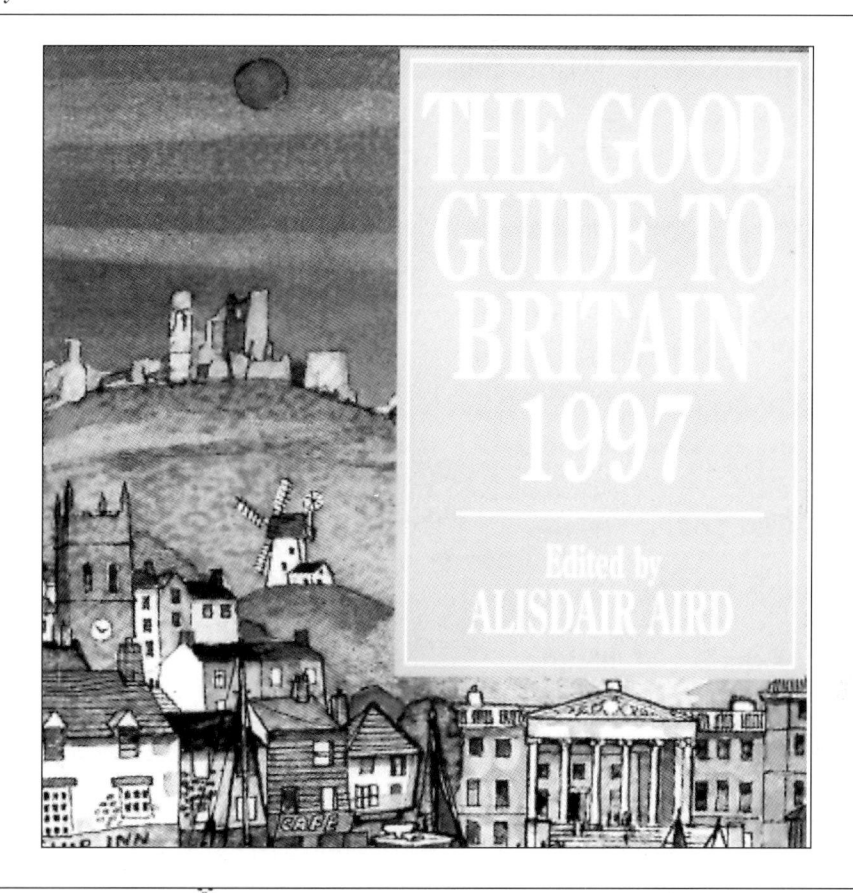

Edited by ALISDAIR AIRD

OXFORDSHIRE FAMILY ATTRACTION OF THE YEAR

Wigginton Heath SP3833 WATERFOWL SANCTUARY AND CHILDREN'S FARM Readers with children very much enjoy visiting this notably friendly family-run farm and rescue centre. One of the things that stands out is the determinedly simple and undeveloped feel – it's perhaps a bit basic for those looking for a glossy full day out, but children between around 3 and 10 love it, and can start cuddling and stroking animals practically as soon as they come through the door. Always plenty of tiny chicks, rabbits, guinea pigs, kittens and the like, especially in the Baby Barn – perhaps the part younger visitors enjoy best. They can feed some of the pigs and ponies, and elsewhere are rare breeds, uncommon aviary birds, and 16 well set out (and carefully fenced off) waterfowl ponds, awash with ducks. There's an outdoor adventure playground, and plenty of space for picnics (some indoors). Concessions for visitors end there, so if you fancy a snack you'll need to nip to one of the farm shops just down the road – or maybe the Pear Tree in Hook Norton. No special events or shows, so the length of a visit depends very much on how long children in your party are likely to want to spend fussing small animals – and how long you don't mind watching them do it. Wear wellies in wet weather. Disabled access; cl 25 Dec; (01608) 730252; *£3.

A sample of the hundreds of visitor comments

Date	Name	Address	Comment
4/5/89	Sue, Andrew, Peter & Richard Wiles	Hillside Close Banbury	Lovely
4/5/89	Nicole, Adam & Sue Boother	Venvell Close, Enstone	Perfect for all the family
4/5/89	Peter & Valerie Wilcock	Broughton Way Banbury 54007	Every wish for success.
4/5/89	Rachel Johnston	The Gables, Sibford Rd, Hook Norton	Much enjoyed
4/5/89	Barbara Rustridge	" " " "	Very enjoyable
4/5/89	Ian & P. Hobday	Sycamore Drive Banbury	
4/5/89	B Holmes	Bloxham	Best wishes
"	M. Nield	Penwortham, Preston, Lancs.	
5/5/89	Queensway Language Resource Centre	Banbury	Lovely, children loved it!
2.8.97	The Skinner Family	Place de l'Eglise, Mersch G-D LUXEMBOURG	Wonderful place. So natural + delightful for children
2.8.97	Victoria + Charlotte	Dorchester on Thames Oxon.	We love our new baby bunnies
4/8/97	Suzannah + Danielle	Heald Green, Cheadle, Cheshire	Wonderful place we loved cuddling all the animals. Thank you
6/8/97	Margarita + Fernando Senar	c/ Modolell, BARCELONA (Spain)	A very nice place with a charming lady
7.8.97	Asumi Matsunaga, Yoshiya, Yuko	Cavell Rd, Oxford originally from Manzaka, Yokohama JAPAN	VERY FRIENDLY Hope to come back soon
11/8/97	ANGUS MARSHALL	MONAL RD, BANGOR 2234 AUSTRALIA	Great for children especially the rabbits!
11/8/97	Sarah Rhian Sturrock	Bryn Farm, Caernarfon, Gwynedd North Wales	Brilliant. Lovely animals and the rabbits are lovely
11/8/97	Richard Sturrock	Bryn Farm, Caernarfon, Gwynedd North Wales	very good animals and Lovely youngsters
13.8.97	Jack Lewis	Horsley road, North Chingford, London E47HX	I had a lovely welcome, it was fun. I enjoyed the animals.
13.8.97	Hannah Lewis	Horsley Rd. North Chingford, London E47HX	I enjoyed the day very much and the animals were all very good.
20.8.97	Beatrice Mancini	Via Fra Galgario, 21100 Varese Italy	It is a wonderful cheerful and loving place, the animals are very well kept, I will come back
20.8.97	Vanessa Mancini	"	Thank you very much. I hope this place will always be the same!
20/8/97	Patrick Kinneir	Hong Kong	Thanks
20/8/97	Walter Kinneir	Hong Kong	Excellent and fun for children of all ages.

Thank you my duck!

Date	Name	Address	Comment
20.8.97	Beatrice Mancini	via Fra Galgario 21100 Varese Italy	It is a wonderful cheerful and loving place the animals are taken very well and I will certainly come back
20.8.97	Vanessa Mancini	"	Thankyou very much I hope this place will allways be the same!
20/8/97	Patrick Kinneir	Hong Kong	thanks
20/8/97	Walter Kinneir	Hong Kong	Excellent and fun for children of all ages.
20/8/97	HOLLY DALBY + FAMILY	LANCASTER	I bought two of Maddie's Rabbits, and I want to come back again.
21/8/97	The Cordingley family	Hong Kong.	Wonderful hands on experience for us all - Thanks!
21st 05 August	Chris Harding	Polstead RD Oxford	I came on my 8th B day
22nd 03 Aug.	Amy and Emma Lloyd	Archery Close London W2 2BE	It is my best farm.
23/8/97	HILDA WHITE	PALING ST BALLARAT. Vic. 3350 AUSTRALIA	AMAZINGLY BEAUTIFUL CONGRATS
25/8/97	Julia A. Jagorskaya	Omsk. Siberia.	I am suprised and enjoyed.
29.8.97	Jasmine Holz	Leythe Road, London	
29.8.97	NAIMA VAN DER BEEK	Corsica FRANCE	very good
29.8.97	Lola Usciolda	Corsica France	c'etait très beau!
2/9/97	DECLAN, MARTINA & LIADAIN EVANS.	8 Lanascar Rd, London	We came yesterday. Loved it so much. We came back today!
" " "	Graeme Marilyn Amy Ellie & Jack.	Cowley Oxford.	visited lots of times love it here very much.
22/9/97	EDWIN LAMBERT	BIDDENHAM BEDFORD	Lovely. Nice to see so many happy birds & animals
22/IX/97	S. G. Colverson (Tom) 01865 37 6052 Publisher	Mill St, Kidlington OXON., OX5 2EE	A splendid venture: we came because we saw 'Welcome to the Cherwell Valley'. We will come again!
?/9/97	DOROTHY EVANS WENDY KYLE	PEMBROKE BERMUDA	Delighted to see such lovely affection
4.10.97	Emma		I held all the kittens
5.10.97	BOBBY, TRUDI BRITTANY & TIERNEY	DAURO.	WHAT A LOVELY DAY THANK YOU X.
10/10/97	Rosa Shimberg	KFAR-SAVA ISRAEL	Beutiful place!

Date	Name	Address	Comment
12/10/97	Tara Kirsty Vowinkel	Schwetzinger Str. 145 Germany	We come every year
17/10/97	Rolf Klint Åke Melander (oldboys)	Stockholm Sweden	a New 'experience' for us! Great fun
18/10/97	S.D. Speake	White Way, Kweglo Gran	Our first visit, it won't be the last.
20/10/97	Fanky Burgess	Cardiff, S. Wales	Best Farm/Park I've ever been too. Thankyou.
21/10/97	Schlrubitz J.H.	1865 Les Diablerets Switzerland	That's very good
27/10/97	Russell Tennent	Belmont Road Uxbridge UB8 2PZ	I liked playing in the adventure playground
29/10/97	Phin J.C.	Mill Street Islip Oxon OX5 2SY	Thanks for an enjoyable visit!
29/10/97	Marc Church	Suncrest Church Close Islip Oxon	Thanks for all the fun.
29/10/97	Alix Campbell	Islip	I want all the kittens.
6.11.97	~~Richard~~ Magda	Mandyla, Water Krakow (POLAND)	Happy animals... as they should be...
27/12/97	Sandi & Terry Wright	The Old School Church Enstone	Wonderful Breakfast
27/12/97	Robin, Karen, Daniel, Leslie, Adam, Gavin, Barry & Richard	Cardigan Place Kettering	Back again from last visit in August
29/12/97	Alanna Winter	Rambler Crescent Beachhaven Auckland New Zealand	I had a lovely time once again.
29/12/97	Sally & Joy Mullins	1 Oxford Road, Horspath, Oxford OX33 1RT	My second home!!
8.1.98	M.K. Henderson	Figtree, Wollongong N.S.W Australia.	
10.1.98	Helen Worrall, Garry Smith, June Sirirak, Mae Thongsuk	10 Bliss Mill Chipping Norton	Lovely personal touch, so welcoming, + nice to go somewhere where children are actively encouraged to touch.

Danny and Thomas misbehaving as usual

The new stockman's cottgage, which secures the Sanctuary for the future.

> Jesus told his disciples about the need to pray continually and never lose heart:
>
> *There was a judge in a certain town who had neither fear of God nor respect for man. In the same town there was a widow who kept on coming to him and saying, "I want justice from you against my enemy!' For a long time he refused, but at last he said to himself, Maybe I have neither fear of God nor respect for man, but since she keeps pestering me I must give this widow her just rights, or she will persist in coming and worry me to death".*
>
> *And the Lord said, 'You notice what the unjust judge has to say? Now will not God see justice done to his chosen who cry to him day and night even when he delays to help them? I promise you, he will see justice done to them, and done speedily......*
>
> Luke 18: 1-8

Cyril

INDEX
Page numbers in *italic* refer to picture captions.

Aaron (from Africa)	49, 56
Abigail	69
Abigail (from London)	101
Adam Stores, Enstone	125
Adderbury Contact magazine	19
Adderbury lakes	19
adoption schemes	21, 26, 31, 39, 41, 89
adults' comments	95, 97, 101, 113, 126, 131
adventure playground	18, 47
Agnes (aunt)	123
Alison (teacher)	109
Allen, Joy	69
Allen, Nick	55, 65, 66, 69
Andrew (photographer)	84
Andrews, Mrs.	13
Andrews, Nigel	9
Andy (headmaster of Neithrop School)	99
"Anglia Television" 'Survival' series	84, 85, 121
Annie	119
anniversaries	
first	41, 46
second	68
Anthony (son-in-law)	6, 45, 50, 76, 96, 99
Arber, Joanne	65
Archie (policeman)	33, 62, 75
Ark Magazine of Rare Breeds	14
Arncott	1
Ashton, Chris	26, 47
Audrey (cousin)	123
aviaries	22, 108, 110, 116, 123, *129*
Axton, Tony	88
Babbs, Sarah	118
Baby Barn extension	*129*, 141
bad-tempered pets	28
Badger, Gill	40
Badger, Jack	4, 39, 40
badges	18, 27
Baldry, Tony 16, 30, 103	
"Banbury Cake"	29, 43, *70*, 75, 89, 117
Banbury Camera Club	54
"Banbury Citizen"	*112*
Banbury Cross Players	141
"Banbury Guardian"	24, 41, 43, 90, 120, *124*
bank manager	31, 45, 94, 107, 120, 134, 137
bantams	40, 52
Frizzle	*58*
Gold Sebright	55
Lucky	54
Pekin	50, 54, 55, 56
Poland miniature	76
Silver Dutch	48

Barlow, Bridget	102, 110
barns	7, 118-19, 120, 128, *129*, 141
Bartlett, Tom	88
Baughan, Nick	13
"Bell", Enstone	33
Ben (from Bodicote)	87
Betty (American)	85
Birdland, Bourton-on-the-Water	35
birds as watchdogs	43
Birkbeck, Alan	14
birthday parties	80
Black Caps	49
blackbirds	47, 48
Blake, Elizabeth	119
Boning, Brian	11, 47, 62
Boswell, Alan and Shirley	131
Bowes, Stan	111
Bowstead, Nancy	120
boy scouts and cubs	45
Boyce, Mr.	22
Brailes	32, 110
breeding programme	19, 55, 99, 102, 116, 118, 140
Brian (policeman)	*129*
Bridget	105
Bridget (from Hook Norton)	88
Brinsea incubators	98, 102, 134
British Waterfowl Association	11, 14, 88, *132*
Broughall, Caroline	92
Broughton Castle	62, 63
Broughton, Suzi	107
Brown, Philip	107
brownies	
Bloxham	125
First Cropredy	94, *94*
First Easington	22
First Sibford	90
Kidlington	54
Bubb, Florence	21
"Buckingham Advertiser & Review"	85, *86*
Buckley, Ray	14, *23*
budgies	62, 94
Bullock, Margaret	116, *116*
bungalow	123, 130, 131, 132, 138, *168*
Butcher, Derrick	90
Buxton, Mike	105
cage size	55
canaries	59
cannibalism	32
Carter, Lorraine	*22*
Carterton	34
cattle	
cow shelter	61
Highland	77, 80
Limousin	118

"Central Television"	25, 103, 144
Charlton-on-Otmoor	1, 4
Cheal, John and Margaret	119
Cherry family	41
Cherry, John	119, 131, 140
Cherry, Martin	114, 119, 120, *129*, 131, 138, 140
Cherry, Mr.	87
Cherwell District Council	20, 38, 69, 103, 132
Chester, Colin and Tina	111, 130
chicks and chickens	67, 94, 119
Andalusian	55
Appenzella Spitzhauben	50, 55, 88
bantam (see bantams)	
Black Orpington	70, 84
Black Pekin	28, 84, 111
Black Rock	119
Blue Cochin	59, 111
Blue Pekin	111
Blue Silky	123
Brahama	41
Buff Cochin	20, 27, 111
Buff Orpington	31, 49, 55, 126, 158
Cochin	41, 111
Columbian Rose Crown	18
Croad Langshan	102
hatching	121, 158
Leghorn	31
Light Sussex	18, 55, 133
Lucy	28
Maran	76, 90, 111
O.E.G. Wyandotte	111
Partridge Cochin	20, 55, 111
Pekin	70
Phoenix	70, 109, 111
Poland	28, 84, 111
Rhode Island Red	55, 89, 90, 111
Sebright	28
Shaver	43
Silky	55
Silver Laced Wyandotte	84, 111
Sumatra Game	27, 31, 39, 55
Sussex	126
Trouser Leg Cochin	28
Warren Shaver	119
Wellsummer	70, 90, 111
children	
comments made by	28, 30, 45, 60, 69, 73, 87, 95, 99, 113, 126
cruel	28, 83
letters from	146-51
terrified of animals	79
Children's Farm	119
Children's Playground	68
Children's sheet	*127*
chipmunks	22, 59, 60, 72, *157*
Dippy	74, *74*

Christopher (from Banbury)	118
Clark, Miss	14
Clarke, Edward and Eleanor	101
cleaning	53
cockatiels	22, 27, 62, 68, 119
Chuckie	48, *48*
cockatoos	119-20
Coles, Trevor	105, 106
colostrum	92
"Cotswold Diamond"	*141*
Community of St Clare, Freeland	156
Cotswold Wild Life Park	111
Cowlan, Mike	68
Cox, Stanley	69
Crackle, Eve	117, *117*
Craig (child from hospital)	21
Crawford, Cyril	4
Crawford, Tony	77
Crofts Pet Shop, Banbury	38
Croughton Cabin	114
Croughton Corner	110, *129*
"Crown and Cushion", Chipping Norton	111
crows	49, 70
Jimbo	87
"Daily Telegraph"	39
dangerous animals, licence to keep	139, *139*
Daniel, Diana and Jonathan	56
David (from Sibford)	53
Deborah (from Deddington)	119
Deddington School Fete	98
Deely, Eva (grandmother)	1
Derek (postman)	126
Dickens family	3
Didcot, group from	126
Dion	119
disabled visitors	14, 35
dogs	
Annie	111
killer Jack Russells	96
Tara	*156*
"Domestic Fowl Trust"	5, 88, 137
Don	123
doves	47
Diamond	35
ring-necked	48, 55
white	109, 130
dragonflies	50
ducks	10, 12, 13, 16, 22, 31, 40, 53, 68, *78*, 81, 95, 118, 158
Appleyard	14, 21, 46, 76, 84
Aylesbury	49
Black Tufted	120
Blue Muscovy	41
Buff Orpington	31, 49, 158
Campbell	18, 65

Carolina	21
Cayuga	17
courting	56
Crested	88
Daffy	27
East Indian	20, 49, 72, 83, 85
Edd	78
Grace Jones	88
Indian Trout Runner	61, 62
Khaki Campell	44
"Love a Duck" badges	18
Lucy	44
Lyson Teal	21
Mallard	19, 21, 27, 51
Mandarin	21
Marianne	92, 93, *93*
Muscovy	12, 18, 20, 22, 26, 27, 30, 40, 44, 56, 130
one-legged	116
Ping Pong	88
Pinny	14
Pintail	21
Pochard	16
Quackers and Quackling	13
Rouen Cross	40
Ruddy Shell	27
Saxony	108
Shelduck	16, 47
shows	88
Silver Appleyard	31, 81
Spats	88
Teal	21
Thomas	56
Trevor	18
Welsh Harlequin	54
White Call	21, 46, 49
White Runner	39, 46
Durndel, Jack	4
Easton, Sue	110
Ebony	49
educational packs and fact sheets	90, 122
eggs, incubation and hatching	*79*, 80, *82*, 90, 158
Ekers, John, & family	22, 73, 110
electricity supply	13-14
Elliston and Cavell, Oxford	4
Emma (from Nuneaton)	56, *56*
emus	139
Enid	119
Enstone	33, 41
Evan (from America)	69, 83, 98
Ewer (Warrant Officer)	114
fencing	20, 43, 51, 70, 139
electric	61, 67
Fencott	1, 2, 4, 52

film shoot	62, 63
finches	59
"Fine Lady Bakeries"	41
flower shops	4, 17, 30, 48, 110
Folly Farm	5
food supply	32, 40, 41, 43, 72, 75, 89, 114, 119, 120, 125, 137
Fowler, Madison Anne	21
foxes	8, 37, 55, 61, 66, 69, 89, 158
Franklin, Andrew	1
Freeman, David	29
"Friends of the Earth"	102, 108, 109, 117
frogs	13, 68, 76, 92
future mission	130
Gaffrey, Mrs.	118
Gardener, Heidi	140
Gardener, Jeanette	36
Garsington Women's Institute	85
geese	10, 28, 37, 83
Beaker	70
Canada	37, 47, 48, 49, *65*, 72
Carolina	47
Chinese	13, 37, 92, *132*
Doris	70
Embden	36, 42
Greylag	130
Ling	*156*
Little Fizzy	100
Mindy	32, 36
Oddsey	70
Sebastopol	19, *20*, *23*, 29, 44, 72, 92, 111, 158
sheep attacked by	36
White Chinese	26
Wiggin	*156*
Zinny	38, 42, 43, 44
"Geoff Reeves Films Limited"	63
Georgina (photographer)	75, 76, 89
Gilkes, Andy	6
Gilkes, Helen	140, *140*
Gillett, Jane	126
Gleadle, Dr. Ron	*132*
goats	9, 16, 20, 26, 34, 37, 43, 51
Angora	53, 158
Billy	90
Bobby	53, *53*
British Alpine	23
British Saanan	9-10
Brock	34, *34*, 51
Cashmere	158
Claire	85
Daniel	23, 30
Danny	42, 43, 50, 51, 67, 107, *167*
feeding	24
Fred	92, 93, *93*, 99
Guernsey	9, 34, 89-90
Poppy	9, 30, 34, 51, 89, 118

Pygmy	9, 34
Roger	53, 85
small-horned	16
Snowy	26, 30, 51, 81
Thomas	9, 16, 30, 44, 51, 67, 76, 86, *167*
wethers	34
Whiffy	34, 51
goldfish	8, 11, 13, 21, 38, 45, 50, 68, 92
Gordon	123
Gosford Hill, Kidlington	4
Grace, Ian	133
Gray, Mrs.	28
Great Rollright	48
grebes	66, 113
green-legged rail	141
greenfinches	75
guide books	47
guinea fowl	43, 47, 98
guinea pigs	27, 28, 31, 53, 56, 89, 130
Hall, Mrs.	36
Hamilton-Gould, Toddy	23, 30
hamsters, Russian	32
Hancox, Darren	17
Handley, Grahame	103, 105, 108
Hanff, Betty	93, 103, 107
Harrap, Sue	126
Harry (photographer)	128
Hart, Michael and Marion	14
Hatcher, Mike	88
Hatwell, Daisy	*3*
Hatwell, Mr. and Mrs.	*3*
Hawes, Chris	*132*
Hawthorn, Jackie	134
Hawtin, Frank	26
healing power, therapy	34, 47
Heather (from Kingham)	123
hedgehogs	110
hedges	51
hens (see chicks and chickens)	
herbal remedies	28, 29, 34, 39
Hethe Victorian Fayre	55
Higham, Stephen	121
Hill, Georgina	128
Holt, John	20, 69, 71
Holt, Joyce	*14*, 17, 19, 21, 69, 71, 72
Homestead Farm, Fencott	1, 2, 4, 52
Honey, Albert	25
Honeybourne, Ben	62, 63
Hook Norton	33, 36
Hook Norton Post Office & Stores	125
Hook Norton Rural Fayre	51, 59, 79
Hope Walker, Vickie	121
Horton-cum-Studley	4, *5*
House, Mr.	128

Hughes, Jan and Roger	59
Hughes, Marcus	47, 68, 130
Hughes, Mr.	16
Hughes, Mrs.	49
hummingbirds	52
Humphries, Wendy	87
hunting	30-1, 60-1
Hutt, Carl	113
ibis	81
inbreeding	90
incubators	93, 98, 102, 118, 134
Invine, Dorothy	*18*
Invine, Roy	7, 7, *9*, 61
Israel	12
Ivings, Ted and Marilyn	41
jackdaws	46, 67, 75, 121
Jack	46
Jacko	75, 80, 84-5, 87, 121
Jacobs, Tony and Pippa	126
Jan (from Hook Norton)	68
Jane (from Marston St. Lawrence)	9
Jean and Peter (from Standlake)	35
Joanna (from London)	33, 35, 43, 73
John (baker)	125
Johnson, Bessie	4
Johnson, George and Ruby	99
Johnson, Mark	67
Johnson, Mr.	134
Jolliffe, Ted and Mary	120
Joy (from Horspath)	126
Joyce (from Charlton)	123
Judy (from Manor Farm)	70
Judy (from Middle Barton)	123
June	13
Kakarekes	19
Kathy (Banbury pet shop)	46
Katie	60
Kay (from Coventry)	130
Kent Adventure Club	131
kestrels	8, 50, 84
Kingsman, Lorraine	118, 119
Kingsman, Victoria	106, 119
labels and nameplates	140
Lacey, Tessa	29
"All about Geese"	23
Lakin, Pat	123
lambs (see sheep and lambs)	
Lay, Trevor	16
leaflets	79, 120, 121
Lee (grandson)	58, 80, *144*
Leek, Tiffany	*100*
Licence, Ben	18

Linda (from Banbury)	118
Lishman, I.M.	*112*
Listening Books for the Blind	22
Little, Reg	62, *63*, 65
Longhope	52
love between goose and goat	43
love-birds	86
lovers' seat	125
Lowther, Mr.	108
Lucy (from York)	94-5
Lyn	26, 29
Lynn, David and Claire (from Brailes)	35
M40 motorway	51
McLellan, Andy and Sharron	17
Madeiros, Clive	139
magpies	41, 49, 70, 92, 94, 112
Asian	88, 92
Malcolm's Bakery, Banbury	32
map of sanctuary	65, *66*
Marcus	67, 101, 111
Margaret	36
Margery	43, 69, 74, *74*
Mark	119, 120, 138, *138*
Martin	119, 128
Matthews, Norman	106
Mel	108
Melanie (of Bloxham)	92
Melody (granddaughter)	*34*, 53, *53*, 88, 95, 116, 128
Micah (grandson)	75, *75*, 95, 106, 116, 128, *138*
mice	38, 45, 54, 74, 101, 110, 113
fruit bat	113, 116, *117*
Michael	67, 123, 128
Middleton Cheney	110
Milcombe Fete	52
Milcombe, New Road Stores	125
Ministry of Agriculture	30
Mobbs, Paul	102, 103, *105*, 106
moorhens	50, 55, 61
Morgan, John	118
Morris, Bob	81, 130
Morton, Ann	118, 128
Morton, David	128
Muldoon, Royston	107
Muriel (from Wantage)	102
myxomatosis	24, 131, 158
National Exhibition Centre	96-7, 98
National Rivers Authority	107
nature trail	37, 77
Newbold-on-Stour Village Fête	84
newts	19
North, Brian	121
"North Cotswold Diamond"	141
North, Darren	14, 18

North family	*2*
North, Mark	138
North, Martin	120
North, Maurice	*9*, 14, 18, *23*, 114, 119, 120, 128, 138, ***138***
North Newington	38
North, Peggy (sister)	*2*, 12, ***12***, 109
North, Sue	121
North, William	*3*
nuthatches	60
observation tower	8, 61
ostriches	139, 140
Owl and Hawk Society	123
owlery	114
owls	34, 52, 56, 66, 67, 68, 76, 100, 101, 123, 125
barn	20, 52, 67, 97, 110, 112
claws	60
eagle	139
little	8
Olitwo	74, 76, 88, 139
Oliver	33
Ollie	58, 59, 60
Owlyn	111
tawny	8, 58, 74, 123, 125
"Oxford Mail"	22, 24, 27, 28, 35, 144
"Oxford Times"	62, 63, 65
Oxfordshire County Council	103, 105, 112
Oxfordshire Special Conservation Award	62, 64-5, ***64***, *71*
Page, Michael	69, 90, 94, 98, 99, 118
Pam (from Milcombe)	125
parakeets	
Cocky	59, ***59***, 125, 139
friendship with rabbit	*59*, 125
ring-necked	59
parrots	
Blue Amazon	5, *5*, 20
Polly Parrot	20
partridges	83, 84
Paul (policeman)	123
Paul (teacher)	99
Paws Pet shop, Banbury	75
Payne, Mary	119
Payne, Richard	110, 119, 132
peacocks	11, 24, 27, 41, 45, 48, 49, 55, 61, 66, 68, 69, 83, 84, 89, 95, 111, 131, 158
Indian Blue	137
Javanese Green	137
Percy	***11***
Pointer	84
White	70, ***71***
Pearce, Catherine	111
Pearce, Geoffrey	98, 134
Pegram family	89
pens	9, 130

Peter (from Middle Barton)	35, 48
Pets Corner	76, 84
pheasants	52, 54, 70, 73
Blue-eared	110
Golden	47, 53
Monal Himalayan	110
Reaves	13
Silver	17, 32, 110
pigeons	16, 31, 49, 58, 76, 89
Birmingham Roller	22, 33
Fantail	47, 49, 57, 76
foster mums to rabbits	88
racing	47, 48, 95
pigs	40, 57, 60, 14
Dora	112
Dotty	112
Gloucester Old Spot	101
housetrained	57
Oxford Sandy and Black	60
pig sitters	57
Vietnamese pot-bellied	57, 112, 114
Winniepiglet	60, 102
Winston	60
plan of sanctuary	*127*
planning permission	9, 117, 130-7
playgroups (see schools and playgroups)	
ponds	10, 19, 35, 39, 48, 50, 55, 131
cleaned and dug out	112, 120
dried up	23
fencing	36
goldfish	8, 11, 13, 21, 38, 45, 50, 68, 92
island	16
tadpoles	76
ponies	67
Penny	54, 67, 89, 92
Vienda	67, 73
Whiskey	67
Potter, Alex	70, **70**, 92, 110
Potter, Cliff	110
predators	37, 49, 118
Price, David	97, 130, 131
Price, Jean	97
Price, Roy	88
puffins	33, 34, 52
puppies	103
rabbits	9, 10, 24, 30, 37, 41, 42, 43, 45, 47, 50, 52, 53, 55-6, 59, 76, 121
Angora	10, 31, **65**, 100, 158
baby	42, 45, 46, 50, 68, 84
bad-tempered	114, 158
breeding programme	44, 114, 140
broken heart, death from	139
Buttons	47
Cashmere	68, 158

Cyril	168
Dwarf Lop-eared	85
Dwarf Netherland	51, 55, 76
female behaviour	157, 158
fostered by pigeons	88
friendship with parakeet	*59*, 125
frustrated	47
George	10, 28, 29, 31
Giant Himalayan	56
Horace and Hilda	46
hypnotized	*34*
Jealous "Jelly"	*29*
Jo	29
Mabel	95
Mabel's Midgets	*42*, 73, 114, 116, *133*, 137, *167*
Mary Poppins	108
Micro Midgets	118
"mini bunnies"	108
myxomatosis	24, 131
New Zealand White	46
Old English	38
rabbit warren	9, 24, 26, 44, 48, 57
Reabie	10
Ruby	*59*, 125, 139
Snowy	74, 125
Snuggles	50
Squirrel	10, 44
three legged	49
vicious	36
"Radio Oxford"	28, 29, 41, 47, 133
Radway, John and Margaret	94, 119
rare breeds	27, 75, *132*, 141
Rare Breeds Survival Centre	5
Ratcliffe, Amy	92
rats	37, 57, 84, 85, 113
reception barn	*16*, 45, 58, 59, 60
Reid, John and Sandra	139
rheas	140
Ridley, Marcus	100, 112
Robbins, Bill (first husband)	4-5, *4*, *5*, 53, 97
Robbins, Jackie (daughter)	*5*, 6, 45, 50, 58, 99
Robbins, Rodney (son)	5,7-8,13,27,30, 33,45,48,52,54,55,59,60,61,62, 64,65,66,68,73,79,81,84-5,87,108, 109,110,111, 114,117,119,121,128,*129*,130,131,138,*138*
Robbins, Wendy (daughter)	43, 45, 50, 51, 52, 53, 59, 62, 66, 67, 68, 70, 73,79, 81, 83, 89, 90, 92, 95, 99, 101, 103, 106, 114, 116, 117, 119, 128, 133
Roberts, Doreen	125
Roberts, Michael	88, 137
Roberts, Tony	*116*
robins	94
Roger and Jan (from Hook Norton)	48
Roy	18, 22, *23*, 47, 57, 125
RSPCA	10, 23, 25, 50, 51, 57, 74, 87, 123
Ryan, Susan	46, 55, 63, 69

Ryan, Victoria	69
Sainsbury's	41
St. Lucia	16, 27
salmonella	30, 43, 49
Sammons, Anne and Tom	125
Sandle, Phyllis	47
Saxton, Terry	130, 133, *133*
Schneider, Mrs.	59
schools and playgroups	
Barley Hill	139
Barn Nursery	23
Berkhamstead	128
Birmingham	131
Bishop's Itchington	118
Blackbird Leys	126
Bletchington	123
Bloxham College	22, 73
Chadlington	123
Charlbury Primary	114
Chatswoth Primary, Hounslow	130
Coventry	109
Cropredy	23
Croughton Middle	96, 103, 107, 108, 110
Croughton USA	114
Cutteslowe	118, 123, 131, 154
Deddington Primary	76
Dr. Radcliffe's, Steeple Aston	77
Emmanuel Christian, Oxford	114, 153
Enstone Church Sunday School	23
Fairway Playgroup	121, 123
Finmere	114, 118, 128
Freeland	76, 81, 83
Glory Farm School, Bicester	80
Great Tew	118
Great Tew Playgroup	20
Greenway First School	155
Greenmere School, Didcot	153
Harnham, Sailsbury	112
Headington	155
Henry Box, Witney	81, 130
Hook Norton	113, 114
Hook Norton "Jack in a box" Nursery	52
Horspath	118, 119, 153
Kidlington Mums & Toddlers Group	81
Little Compton	98
Lois Weedon Playgroup	130
Longfield, Bicester	130
Middle Barton	95
Middleton Cheney	80
Mollington	155
Neithrop	45, 97, 99, 110
North Hinksey	114
Overthorpe	107, 108, 113
Penhurst, Chipping Norton	*18*

Queensway	76
St. Ebbe's, Oxford	77, 154
St. John's, Banbury	19, 23
Ss. Philip & James, Oxford	154
Southfield, Brackley	81
Stadhampton	128
Standlake	126, *126*
Steeple Aston	25
Steventon Playgroup	80
Tysoe	154
Warriner	32
Witney Sunday School	47
Woodeaton Manor	76
Wooton	123
Scott, Sir Peter	29
seagulls	65
seats	125
sheep and lambs	47, 69, 70
Cotswold	81, 89
dipping	61
Jacob	90, 125
Laddie	70
Larryadne	70, 90
Lindsie	70
pens	125
pet lambs	14
Ruth	83
Sarah	90, 125
Soay	10, 26, 34, 51, 70, 76, 125
Tex	92, 93, *93*
Wiltshire Horn	140
Zeb	26, 51, 76
Zoë	26, 111
Shell, Mr. and Mrs.	130
Shepherd, Tim	85
Shipston-on-Stour, Rural Fayre	74, 97
signs	14, 30, 38, 46, 59, 67-8, 125, *145*
Simms, John	132
Slimbridge	5, 39, 130, 131
Smith, Bernard	19
Smith, Gordon	107
Smith, Mark	52
Smith, Mr. and Mrs.	119
Smith Ryland, Mr.	30
Soaffe, Joyce	13
Society of Friends, Sibford	120
Soden, Philippa	*72*
sparrowhawks	67, 84
sparrows	61, 76
squirrels	60, 61
Stephanie (granddaughter)	106, *144*
stoats	11, 21, 22, 27, 42, 45, 49, 54, 57, 81, 83, 84, 85
Stockford, Colin and Madge	*129*
Stoke Mandeville Hospital	93
Stoneleigh	133

Stowe Fair	73
Stuart	44
Swalcliffe Fête	50, 81
swallows	49
Swan Foundry, Banbury	31
"Swan" guest house, Enstone	92
"Swan Lake"	107
swans	*11*, 87
black	11, 14, 42, 44, 52, 117
Coscoroba	62, 66, *66*
Diane	62
Samantha	24, *24*, 25, *25*, 27, 29, 36, 56, 107, 110
Swan Watch	87
Tarrant, Fiona	144
taxidermy	16, 20, 34, 52, 60, 113
Taylor, Bernard	69
Taylor, Charles (father)	1, *2*, *3*, 4
Taylor, Christopher (brother)	1, *2*
Taylor, Eileen	11
Taylor, Eva (grandmother)	*3*
Taylor family	*3*
Taylor, Jabus (great-grandfather)	1
Taylor, John (brother)	11
Taylor, Jonas (grandfather)	1, *1*, *3*
Taylor, Lucy Lockett	36
Taylor, Mary (mother)	*2*, 3, *3*, 4, 109
Taylor (North), May (sister)	*2*, *3*, 17, 46
Taylor, Raymond (brother)	*2*
Taylor, Ros	88
Taylor, Sue	69, 98
teachers, letters from	152
telepathy	45, 93
telephone connection	14
Tennyson, Lucy	22, 24
terrapins	27, 36, 45, 50, 57, 68, 85, 131
Terry	133
theft of animals	61
therapy	34, 47
"Three Pigeons", Banbury	57
thrushes	84
ticks	31, 42, 43
Tim (photographer)	84
"Times Educational Supplement"	96, 101, 102
toilets	125
tortoises	111
Fred	94
Tourist Boards	14, 41, 46, 47, 111, 133
trade, license to	20
Treadwell, Bernard	*3*
tree planting	45
Trifleton Waterfowl Collection	6
Tudor Photography	50, 128
turkeys	19, 31, 35, 66, 68
Turner, Colin	134

Tustain, Becky	27
Tustain, Mr.	33
Underwood, John	27
Valdambrini, Richard	121, 128, 130, 133
vets	
Clive Madeiros	60, 74, 139
David Shepherd	111, 130
Kate	80, 84, 111, 137
Tony Roberts	*116*, 139
West Bar Hospital	56, 66, 80, 84, 111, 125
video promotion	121, 128
Vine, Penny	18
Wade, Sara	118
Walker, Mr.	96
Ward, Charlie	125
Warner, Dennis (second husband)	6-8, *6*, 7, 47, *72*, 87, 97, 109, 114, 119, 133
Warner, Mabel	*2*, *5*, 7
childhood	4-5, 51-2
family	1-3
marriages	4, *4*, 6, *6*
Warren, Joan	130
Warren, Judy	119
waste tip	102-9, 112, 117
water supply	54-5
Waveney Wildfowl	16
weasels	49, 83, 84
Weir, John	105, 106, 108
Went, Walter	85
Wigginton	6
Wilcox, David	60
Williams, Mrs.	19
Williams, Peter	105
Winson, Terry 50,	51
Winter, Alanna	*79*
woodpeckers	100
Wooley, Kate	131
worksheets	*91*, *127*
Wrench, Mrs.	13
yellowhammers	50, 61
zebra finches	40